SpringerBriefs in Computer Science

For further volumes:
http://www.springer.com/series/10028

Ke Xu · Yifeng Zhong · Huan He

Internet Resource Pricing Models

 Springer

Ke Xu
Yifeng Zhong
Department of Computer Science
Tsinghua University
Beijing
People's Republic of China

Huan He
Global Technology Service
IBM North China
Beijing
People's Republic of China

ISSN 2191-5768   ISSN 2191-5776   (electronic)
ISBN 978-1-4614-8408-0   ISBN 978-1-4614-8409-7   (eBook)
DOI 10.1007/978-1-4614-8409-7
Springer New York Heidelberg Dordrecht London

Library of Congress Control Number: 2013942996

Printed on acid-free paper

Springer is part of Springer Science+Business Media (www.springer.com)

Preface

The rapid Internet growth in recent years has seen a dramatic increase in Internet resources consumption. Accordingly, the Internet resource pricing strategy is attracting more and more attention, since it is not only the key factor for efficient Internet resource allocation, but also for the determinant of profits. However, the inefficiency of some existing pricing strategies is, to some extent, impeding the sound and sustainable development of the Internet. In the research field of resource pricing, a whole pricing strategy can be formulated to contain pricing models, service mechanisms, and pricing methods, which cover all related topics. We first introduce three basic Internet resource pricing models through Internet cost analysis and survey several corresponding mechanisms that can ensure pricing implementation and resource allocation. On network resource pricing methods, we discuss utility optimization and emphasize two classes of pricing methods, including system optimization and strategic optimization of network participants. Then, we summarize and analyze the pricing and management problems and the corresponding research work in P2P and mobile networks. Finally, we propose two examples of new pricing strategies to solve the profit distribution problem brought about by P2P free-riding and improve the pricing efficiency in the mobile market for multi-interface users. We hope the models and applications introduced in this book will help to give insights into the pricing of Internet resources.

Acknowledgments

We thank Prof. Sherman Shen for his great help in publishing this book. We also thank Ms. Yingjie Zhang and Ms. Fang Liu for their precious efforts in editing.

This work was supported by NSFC Project (Grant Nos. 61170292, 60970104), 973 Project of China (Grant Nos. 2009CB320501, 2012CB315803), and the National Science and Technology Major Project of the Ministry of Science and Technology of China (Grant No. 2012ZX03005001-001).

Contents

1   Introduction   1
    1.1   Background   1
    1.2   Internet Resource Pricing   2
    1.3   Classification and Comparison of Pricing Strategies   4
    1.4   Organization   6
    References   6

2   Brief History of Pricing Model   11
    2.1   Basic Pricing Models   11
    2.2   Pricing Mechanisms   14
    2.3   Pricing Methods   18
        2.3.1   Pricing Based on NUM   18
        2.3.2   Pricing Based on Game Theory   20
    2.4   Summary   24
    References   25

3   Pricing and Management Related to P2P and Mobile Internet   29
    3.1   P2P's Impact on Cooperation-Based Internet Pricing
        and Profit Distribution   29
        3.1.1   Development of P2P   30
        3.1.2   Charging and Profit Division in P2P Networks   32
    3.2   Pricing Mechanisms in Mobile Internet   35
    3.3   Summary   39
    References   40

4   Cooperative Game-Based Pricing and Profit Distribution
    in P2P Markets   43
    4.1   Non-cooperative Game Model   43
        4.1.1   Network Model   43
        4.1.2   Strategy-Chosen Game   45
        4.1.3   Two-Stage Price-Decision Game   46
        4.1.4   P2P-Involved Profit Computing Model   50
        4.1.5   Examples and Analysis   53

4.2   Cooperative Profit Distribution Model   56
         4.2.1   Profit Distribution Between ISP Coalition
                    and PCP Coalition.   56
         4.2.2   Profit Distribution Within Each Coalition   60
   4.3   Summary   64
   References   64

5 **Pricing in Multi-Interface Wireless Communication Markets**   67
   5.1   Background   67
   5.2   Modeling   68
         5.2.1   Internet Service Market   69
         5.2.2   ISPs' Service Composition Model.   70
         5.2.3   Multi-Interface Mobile Host User Model   71
   5.3   Analysis of the Dynamic Game Process   72
         5.3.1   Payment Computing of Multi-Interface Mobile Hosts . . .   72
         5.3.2   Computing of ISP Profits   72
         5.3.3   Analysis of the Dynamic Game Process in Two
                    Types of Markets   74
   5.4   Simulation and Analysis.   79
         5.4.1   Survey on User Groups   79
         5.4.2   Analysis of the Exclusive Monopoly Market   81
         5.4.3   Analysis of the Oligopoly Market   83
   5.5   Summary   86
   References   86

Chapter 1
Introduction

1.1 Background

Internet resources pricing has attracted more and more attention, because it is not only a key factor to efficiently allocate Internet resources, but also the determinant of the profit. On the one hand, Internet Service Providers (ISPs) make their routing and peering decisions based on interconnect costs and backhaul cost evaluation, so as to reduce traffic cost in their networks [56]; on the other hand, they take efficient pricing strategies to increase network operation revenues. Of course, the other important goal of pricing strategies is to efficiently allocate Internet resources.

Too many packets will incur network performance degradation, which is called congestion [29]. It is caused by unbalanced resources and traffic distribution, and thus will not be automatically eliminated with the increase of network capacity. In a packet switched network, the selfishness of users makes this happen. As illustrated by Hardin [37], "tragedy of commons" occurs when many individuals share public resources and each one holds a selfish purpose, which means the loss they bring to others is larger than the extra benefits they gain. So, if networks serve as public goods, the overall excessive personal usage will possibly cause system performance decline and consequently the congestion problem.

In recent years, with the fast development of QoS-aware video, audio and other bandwidth-consuming applications, network traffic has surged, which makes network congestion more frequent and serious. Accordingly, compared with simple priority-based QoS mechanisms [16, 17], novel content distribution technologies and multi-layer QoS mechanisms have been studied and improved. For the former, a new layer of network architecture, i.e., the application layer network, is added to the existing Internet, such as P2P (Peer-to-Peer [4]) and CDN (Content Delivery Networks [71]). For the latter, commonly, QoS mechanisms are developed to work at multiple layers of a network, such as the transport and network layers, which are widely concerned with basic network service mechanisms, e.g., passive congestion control [48, 55] and traffic engineering [75].

K. Xu et al., *Internet Resource Pricing Models*,
SpringerBriefs in Computer Science, DOI: 10.1007/978-1-4614-8409-7_1,
© The Author(s) 2014

However, network management will be increasingly difficult due to the following reasons. (1) Video-like traffic will keep increasing [12], which indicates higher QoS requirements; (2) For different CDN/P2P applications or ISPs, their selfish QoS control objectives may lead to conflicting behaviors which may even degrade network performance; (3) For multi-layer QoS mechanisms, since they often complicate network protocol design and implementation, the effectiveness is limited. Moreover, as they do not differentiate high-level application types, service-based QoS differentiation is hard to achieve. For ISPs, a convenient way to improve QoS might be network infrastructure upgrade or network capacity increase. However, such short-term investments usually lead to high costs and fail to satisfy the fast-growing network resource requirements in the long run.

In addition, a bold attempt should be mentioned, i.e., proposing network architectures to ensure QoS. For example, IntServ [76] guarantees QoS per-flow resource reservation, and DiffServ [10] modifies IntServ architecture by adding priorities based on aggregated flow. Theoretically, they can improve network efficiency, indicating that a QoS guaranteed era is coming. However, in addition to technical complexity, QoS-guaranteed high priority services are actually achieved at the expense of low-priority ones. Furthermore, as the Internet management is distributed, ISPs lack adequate incentives to collaboratively improve network performance/efficiency. These largely impede the implementation of such architectures.

Then, we can conclude that at the premises of limited resources and partially QoS-aware services, an equally important problem of QoS improvement is how to effectively and reasonably use network resources. As for deploying new architectures, proper incentive mechanisms should be designed as a necessary support.

1.2 Internet Resource Pricing

From the above discussion, we notice that designing incentives at economical levels can guide users to rationally use resources and encourage ISPs to improve network performance, so as to be of great significance in effective network resource management and distribution [13]. Therefore, the key issue is resource pricing as an active resource management method may affect revenue sharing of ISPs. Pricing that provides economic incentives to suit services is of particular importance as an auxiliary for technological progress.

How then do we realize such pricing and what are the key challenges? To answer these questions, three problems should be considered:

Q1. Basically, which factor should be charged?
Q2. How can we identify these factors in different service mechanisms?
Q3. How much should be charged?

As shown in Fig. 1.1, a complete picture of network pricing is presented, including three aspects: basic pricing models for Q1, mechanisms to ensure pricing implementation for Q2, and methods to determine pricing levels for Q3. After we decide the

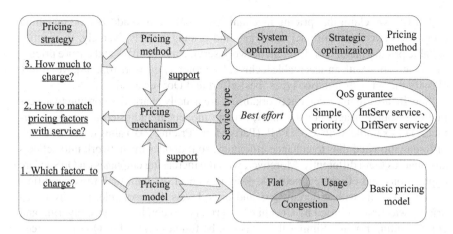

Fig. 1.1 The structure of pricing strategies

pricing factors for specific services and the corresponding pricing methods, a relatively complete pricing is proposed. However, obviously, the computational and technical complexities should be measured before we adopt and implement such pricing. We will briefly introduce each aspect as follows.

For Q1, we define that pricing models decide which factors to charge, or how to evaluate network operation and maintenance costs. Mason and Varian [51, 53] classified the cost as a fixed cost due to the basic service structure (such as leased lines, equipment maintenance, and human resources), marginal costs of accesses, network expansion costs, marginal costs of sending data packets into the congested network, and social costs caused by negative impact on other users. They believe a good price should reflect these costs. Hereby, we introduce three basic pricing models: flat pricing [51], usage pricing [20, 25, 38] and congestion pricing [19, 30, 40–43, 51, 52, 54, 78].

Historically, when applications were simple and resources were sufficient at the early stage of the Internet, it was convenient to charge users fixed fees with a flat pricing model. However, due to the increase of network content, network resource shortage has probably been caused by an excessive number of packets. Then, due to the lack of incentives for efficient network resource usage [26] (a lot of bandwidths are wasted by non-critical applications), the overall network performance underwent degradation. For users, the experience deteriorated and the fairness could not be guaranteed. Thus, flat pricing was no longer applicable. Then, a more effective resource pricing model "usage-based pricing" was proposed [25]. It pointed out that if the charge was usage-based, a fair and efficient use of resources would be moderately promoted to some extent. However, with a continuous increase in network traffic, the aggravated congestion makes the related pricing a hot research area, resulting in a relatively dynamic pricing model "congestion pricing" [51, 53] which has been studied extensively. Besides, the combined use of these three pricing models could be applied because they reflected different cost components.

For Q2, we claim that pricing mechanisms mainly aim to address the matching problem between network service types and pricing models. Namely, to identify suitable pricing factors for different network service mechanisms and ensure pricing implementation with an acceptable technical complexity measure [63, 68], we simply classify the services into two types: best-effort and QoS-enabled.

Specifically, in the former network, users are usually charged according to access rate or resource usage. For the latter, pricing models are adjusted to changed services. For example, Odlyzko's PMP (Paris Metro Pricing [60]) aims to achieve QoS differentiation and thus to enhance efficiency, so it divides the network into subnets and charges them differently. Moreover, with the increasing emphasis on QoS-aware applications and efficiency, network designers and ISPs both tend to serve different data streams with different QoS and price levels. For example, priority-based pricing was first proposed by Cocchi et al. [16, 17] to conduct service layering and corresponding pricing. Similar thoughts can be found in [59]. For QoS guaranteed network architectures (e.g., IntServ and DiffServ), the corresponding pricing mechanisms have been widely studied [14, 21–23, 27, 28, 31, 32, 39, 65, 74, 73, 80, 79].

For Q3, we emphasize how to set a reasonable price level if pricing factors are identified in a specific service. In most cases, prices are the results of supply-demand interactions or competitions. To this end, we will introduce various pricing methods mainly based on optimization theory and game theory. There are two major research paths:

(1) System optimization, i.e., the NUM (Network Utility Maximization [31, 32]) framework, which is largely based on optimization theory [7];
(2) Strategic optimization of network participators, which is based on non–cooperative games [8, 58] (e.g., models in [1, 5, 6, 66, 67, 69]), and cooperative games [8, 57, 64] (e.g., models in [9, 49, 50]).

1.3 Classification and Comparison of Pricing Strategies

As shown in Fig. 1.1, the pricing strategy involves pricing methods, pricing mechanisms and pricing models. In this section, based on pricing models, service mechanisms and price level setting methods, we conduct classification and comparison of the introduced typical pricing strategies shown in Table 1.1. In order to describe the pricing for QoS guaranteed services, we add QoS contract to the pricing model, which represents the achieved services and price agreements between ISPs and users.

In Table 1.1, it should be noted that early pricing models lack theoretical basis, and most of them are based on experiments. So they cannot cover complete decompositions. Some articles focus on studying pricing methods with no differentiation on QoS, so we generally assume they are applicable to best-effort networks in Table 1.1. In addition, the QoS guaranteed types of services refer to what we have described

Table 1.1 Classification of pricing strategies

Pricing model				Service type				Pricing method			Example
Access	Usage	Con[a]	QoS	Best effort	QoS-guarantee			NUM	Game model		
					Priority	IntServ	DiffServ		Non[a]	Co[a]	
	✓			✓							[20, 25]
	✓			✓				✓			[18, 38]
✓	✓			✓							[3]
✓	✓			✓					✓		[72]
✓	✓			✓				✓			[53]
	✓	✓						✓			[40, 41, 61]
	✓									✓	[52] (MD)
		✓	✓					✓			[42]
		✓	✓								[13, 15]
✓			✓					✓			[60]
	✓	✓			✓			✓			[24]
	✓				✓			✓			[16, 17, 61]
		✓			✓			✓			[31, 32]
			✓			✓					[14, 39]
	✓	✓				✓		✓			[28] (MD)
	✓	✓					✓	✓			[2, 65] (MD)
		✓	✓				✓	✓			[73]
		✓				✓	✓	✓			[47]
		✓	✓					✓			[44]
	✓	✓	✓					✓			[10, 36, 35, 45, 62, 70]
✓	✓		✓					✓			[34]
	✓		✓					✓			[33, 46]
	✓		✓							✓	[1, 5, 6, 66, 67]
	✓	✓	✓							✓	[9, 77]
		✓	✓							✓	[49, 50]

In this table, the symbol ✓ in each row represents a feature hold by the pricing strategy example in the last column, and the symbol (MD) means mechanism design
[a] 'Con' stands for congestion pricing. 'Non' and 'Co' stand for non-cooperative game and cooperative game, respectively

in Sect. 2.2. For pricing models, if both usage and accesses are chosen, it means the pricing model is combined with both of them.

Obviously, pricing for different service types inherently has different technical complexities. Generally, for best-effort networks, pricing is always done at the edge of networks, and incurs a lower overhead cost; while for QoS guaranteed services, since pricing relates to QoS along the whole serving path, it involves a higher audition and accounting cost to achieve higher network efficiency and better performance.

In order to illustrate how different ingredients can be combined to build a whole pricing strategy, we analyze several examples here. For example, in [72], Wang et al. studied pricing in best-effort networks by using the flat and usage-based pricing model according to two-player non-cooperative games; while in [2], Altman et al. studied pricing of differentiated services and its impact on the choice of service priority at equilibrium based on non-cooperative games. Especially, for pricing that is based on Shapley value in the cooperative game model, marginal contribution is the only measurement for payoffs, so we leave out such work in Table 1.1.

From Table 1.1, we can conclude that due to the increasing complexity of the service and market environment, the corresponding pricing factors and methods will become even more complicated. Also, we present the evolution process of pricing strategies. It is a manifest trend that when multi-ISP and multi-CP are involved, the game model will be a more suitable and attractive choice. More detailed analyses and applications will be referred in Chaps. 2, 4 and 5.

1.4 Organization

The next two chapters of this book is organized as follows. Chapter 2 presents three basic pricing models, pricing mechanisms based on two types of services, and introduces pricing methods based on two classes of optimization, including system optimization and strategic optimization of network participants in different network marketing environments. Then, we classify and compare typical pricing strategies according to pricing models, serving mechanisms and pricing methods involved. Chapter 3 summarizes and analyzes the pricing and management problems and corresponding research work in P2P and mobile networks.

The last two chapters provide two applications in the Internet access market. Chapter 4 presents a cooperative game-based pricing model and a profit distribution mechanism among ISPs and CPs in the P2P market. Chapter 5 proposes a bargaining based pricing model and analyzes the dynamic game based competition among ISPs in the mobile market of multi-interface users. We hope that these two applications will shed light on the pricing and resource allocation based on game theory.

References

1. Acemoglu, D., Ozdaglar, A.: Competition and efficiency in congested markets. Math. Oper. Res. 32(1), 1–31 (2007)
2. Altman, E., Barman, D., Azouzi, R.E., Ros, D., Tuffin, B.: Pricing differentiated services: a game-theoretic approach. Comput. Netw. 50(7), 982–1002 (2006)
3. Altmann, J., Chu, K.: How to charge for network services–flat-rate or usage-based? Comput. Netw. 36(5), 519–531 (2001)
4. Androutsellis-Theotokis, S., Spinellis, D.: A survey of peer-to-peer content distribution technologies. ACM Comput. Surv. 36(4), 335–371 (2004)

5. Basar, T., Srikant, R.: Revenue-maximizing pricing and capacity expansion in a many-users regime. In: Proceedings of INFOCOM 2002, vol. 1, pp. 294–301. IEEE (2002).
6. Basar, T., Srikant, R.: A Stackelberg network game with a large number of followers. J. Optim. Theory Appl. **115**(3), 479–490 (2002)
7. Bertsekas, D.P.: Nonlinear Programming, 2nd edn. Athena Scientific, Belmont (1999)
8. Cao, X.: Preference functions and bargaining solutions. In: Proceedings of IEEE Conference on Decision and Control 1982, vol. 21, pp. 164–171. IEEE (1982).
9. Cao, X.R., Shen, H.X., Milito, R., Wirth, P.: Internet pricing with a game theoretical approach: concepts and examples. IEEE/ACM Trans. Netw. **10**(2), 208–216 (2002)
10. Carlson, M., Weiss, W., Blake, S., Wang, Z., Black, D., Davies, E.: An architecture for differentiated services. Request for Comments (RFC) 2475 (1998).
11. Chiang, M., Zhang, S., Hande, P.: Distributed rate allocation for inelastic flows: optimization frameworks, optimality conditions, and optimal algorithms. In: Proceedings of INFOCOM 2005, vol. 4, pp. 2679–2690. IEEE (2005).
12. Cisco VNI Service Adoption Forecast, 2011–2016. http://www.cisco.com/en/US/solutions/collateral/ns341/ns525/ns537/ns705/ns1186/Cisco_VNI_SA_Forecast_WP.html
13. Clark, D.: A model for cost allocation and pricing in the Internet. J. Electron. Publ. 1 (1995).
14. Clark, D.: Combining sender and receiver payments in the Internet. In: Interconnection and the Internet: Selected Papers From the 1996 Telecommunications Policy Research Conference, vol. 14, pp. 95–112, Lawrence Erlbaum (1997).
15. Clark, D.: Internet cost allocation and pricing. In: McKnight, L., Bailey, J. (eds.) Internet Economics, pp. 215–252. MIT Press, Cambridge (1997)
16. Cocchi, R., Estrin, D., Shenker, S., Zhang, L.: A study of priority pricing in multiple service class networks. In: ACM SIGCOMM Computer Communication Review, vol. 21, pp. 123–130. ACM (1991).
17. Cocchi, R., Shenker, S., Estrin, D., Zhang, L.: Pricing in computer networks: motivation, formulation, and example. IEEE/ACM Trans. Netw. **1**(6), 614–627 (1993)
18. Courcoubetis, C., Stamoulis, G.D., Manolakis, C., Kelly, F.P.: An intelligent agent for optimizing QoS-for-money in priced ABR connections. http://citeseerx.ist.psu.edu/viewdoc/download?doi=10.1.1.22.291&rep=rep1&type=pdf (1998)
19. Crowcroft, J., Oechslin, P.: Differentiated end-to-end Internet services using a weighted proportional fair sharing TCP. ACM SIGCOMM Comput. Commun. Rev. **28**(3), 53–69 (1998)
20. Currence, M., Kurzon, A., Smud, D., Trłas, L.: A causal analysis of usage-based billing on IP networks. http://citeseerx.ist.psu.edu/viewdoc/download?doi=10.1.1.41.2035&rep=rep1&type=pdf (2000)
21. Dahshan, M.H., Verma, P.K.: Resource based pricing framework for integrated services networks. J. Netw. **2**(3), 36–45 (2007)
22. Dovrolis, C., Ramanathan, P.: A case for relative differentiated services and the proportional differentiation model. IEEE Network **13**(5), 26–34 (1999)
23. Dovrolis, C., Stiliadis, D., Ramanathan, P.: Proportional differentiated services: delay differentiation and packet scheduling. IEEE/ACM Trans. Netw. **10**(1), 12–26 (2002)
24. Dube, P., Borkar, V.S., Manjunath, D.: Differential join prices for parallel queues: social optimality, dynamic pricing algorithms and application to Internet pricing. In: Proceedings of INFOCOM 2002, pp. 276–283. IEEE (2002).
25. Edell, R.J., McKeown, N., Varaiya, P.P.: Billing users and pricing for TCP. IEEE J. Sel. Areas Commun. **13**(7), 1162–1175 (1995)
26. Edell, R., Varaiya, P.: Providing Internet access: what we learn from index. IEEE Network **13**(5), 18–25 (1999)
27. Fankhauser, G., Plattner, B.: Diffserv bandwidth brokers as mini-markets. In: Workshop on Internet Service Quality Economics, MIT, US. Citeseer (1999).
28. Fankhauser, G., Stiller, B., Vogtli, C., Plattner, B.: Reservation-based charging in an integrated services network. In: 4th INFORMS Telecommunications Conference, Boca Raton, Florida, USA, vol. 302, pp. 305–309. Citeseer (1998).

29. Floyd, S., Fall, K.: Promoting the use of end-to-end congestion control in the Internet. IEEE/ACM Trans. Netw. **7**(4), 458–472 (1999)
30. Gibbens, R.J., Kelly, F.P.: Resource pricing and the evolution of congestion control. Automatica **35**(12), 1969–1985 (1999)
31. Gupta, A., Stahl, D.O., Whinston, A.B.: A priority pricing approach to manage multi-service class networks in real-time. J. Electr. Publ. 1 (1995).
32. Gupta, A., Stahl, D.O., Whinston, A.B.: An economic approach to networked computing with priority classes. J. Organ. Comput. Electron. Commerce **6**(1), 71–95 (1996)
33. Hande, P., Chiang, M., Calderbank, R., Rangan, S.: Network pricing and rate allocation with content provider participation. In: Proceedings of INFOCOM 2009, pp. 990–998. IEEE (2009).
34. Hande, P., Chiang, M., Calderbank, R., Zhang, J.: Pricing under constraints in access networks: revenue maximization and congestion management. In: Proceedings of INFOCOM 2010, pp. 1–9. IEEE (2010).
35. Hande, P., Shengyu, Z., Mung, C.: Distributed rate allocation for inelastic flows. IEEE/ACM Trans. Netw. **15**(6), 1240–1253 (2007)
36. Hande, P., Rangan, S., Chiang, M., Wu, X.: Distributed uplink power control for optimal sir assignment in cellular data networks. IEEE/ACM Trans. Netw. **16**(6), 1420–1433 (2008)
37. Hardin, G.: The tragedy of the commons. Science **162**(3859), 1243–1248 (1968)
38. Honig, M.L., Steiglitz, K.: Usage-based pricing of packet data generated by a heterogeneous user population. In: Proceedings of INFOCOM 1995, pp. 867–874. IEEE (1995).
39. Karsten, M., Schmitt, J., Wolf, L., Steinmetz, R.: An embedded charging approach for RSVP. In: Proceedings of the International Workshop on Quality of Service 1998, pp. 91–100. IEEE (1998).
40. Kelly, F.: Charging and rate control for elastic traffic. Eur. Trans. Telecommun. **8**(1), 33–37 (1997)
41. Kelly, F.P., Maulloo, A.K., Tan, D.K.H.: Rate control for communication networks: shadow prices, proportional fairness and stability. J. Oper. Res. Soc. **49**(3), 237–252 (1998)
42. Keon, N.: A new pricing model for competitive telecommunications services using congestion discounts. INFORMS J. Comput. **17**(2), 248–262 (2005)
43. Kunniyur, S., Srikant, R.: End-to-end congestion control schemes: utility functions, random losses and ECN marks. IEEE/ACM Trans. Netw. **11**(5), 689–702 (2003)
44. La, R.J., Anantharam, V.: Utility-based rate control in the Internet for elastic traffic. IEEE/ACM Trans. Netw. **10**(2), 272–286 (2002)
45. Lee, J.W., Mazumdar, R.R., Shroff, N.B.: Non-convex optimization and rate control for multi-class services in the Internet. IEEE/ACM Trans. Netw. **13**(4), 827–840 (2005)
46. Li, S., Huang, J., Li, S.Y.R.: Revenue maximization for communication networks with usage-based pricing. In: Proceedings of the Global Telecommunications Conference 2009, pp. 1–6. IEEE (2009).
47. Li, T., Iraqi, Y., Boutaba, R.: Pricing and admission control for QoS-enabled Internet. Comput. Netw. **46**(1), 87–110 (2004)
48. Low, S.H.: A duality model of TCP and queue management algorithms. IEEE/ACM Trans. Netw. **11**(4), 525–536 (2003)
49. Ma, R.T.B., Chiu, D., Lui, J., Misra, V., Rubenstein, D.: Interconnecting eyeballs to content: a Shapley value perspective on isp peering and settlement. In: Proceedings of the 3rd International Workshop on Economics of Networked Systems, pp. 61–66. ACM (2008).
50. Ma, R.T.B., Chiu, D.M., Lui, J.C.S., Misra, V., Rubenstein, D.: On cooperative settlement between content, transit, and eyeball Internet service providers. IEEE/ACM Trans. Netw. **19**(3), 802–815 (2011)
51. MacKie-Mason, J.K., Varian, H.R.: Pricing the Internet. http://people.ischool.berkeley.edu/hal/Papers/UM/Pricing_the_Internet.pdf
52. MacKie-Mason, J.: A smart market for resource reservation in a multiple quality of service information network. Available at SSRN 975871 (1997).
53. MacKie-Mason, J.K., Varian, H.R.: Pricing congestible network resources. IEEE J. Sel. Areas Commun. **13**(7), 1141–1149 (1995)

54. Milgrom, P.: Putting Auction Theory to Work. Cambridge University Press, Cambridge (2004)
55. Mo, J., Walrand, J.: Fair end-to-end window-based congestion control. IEEE/ACM Trans. Netw. **8**(5), 556–567 (2000)
56. Motiwala, M., Dhamdhere, A., Feamster, N., Lakhina, A.: Towards a cost model for network traffic. ACM SIGCOMM Comput. Commun. Rev. **42**(1), 54–60 (2012)
57. Nash Jr, J.F.: The bargaining problem. Econometrica: J. Econ. Soc. **18**, 155–162 (1950)
58. Nash, J.: Non-cooperative games. Ann. Math. **54**(2), 286–295 (1951)
59. Nichols, K., Jacobson, V., Zhang, L.: A two-bit differentiated services architecture for the Internet. Request for Comments (RFC) 2638 (1999).
60. Odlyzko, A.: Paris metro pricing for the Internet. In: Proceedings of the 1st ACM Conference on Electronic Commerce, vol. 3, pp. 140–147. Citeseer (1999).
61. O'Donnell, A.J., Sethu, H.: Congestion control, differentiated services, and efficient capacity management through a novel pricing strategy. Comput. Commun. **26**(13), 1457–1469 (2003)
62. Ozdaglar, A., Srikant, R.: Incentives and pricing in communication networks. In: Nisan, N., Roughgarden, T., Tardos, E., Vazirani, V. (eds.) Algorithmic Game Theory, pp. 571–591. Cambridge University Press, Cambridge (2007)
63. Roberts, J.W.: Quality of service guarantees and charging in multi-service networks. IEICE Trans. Commun. **81**, 824–831 (1998)
64. Roth, A.E.: The Shapley Value: Essays in Honor of Lloyd S. Shapley. Cambridge University Press, Cambridge (1988)
65. Semret, N., Liao, R.R.F., Campbell, A.T., Lazar, A.A.: Pricing, provisioning and peering: dynamic markets for differentiated Internet services and implications for network interconnections. IEEE J. Sel. Areas Commun. **18**(12), 2499–2513 (2000)
66. Shen, H., Basar, T.: Differentiated Internet pricing using a hierarchical network game model. In: Proceedings of the American Control Conference 2004, vol. 3, pp. 2322–2327. IEEE (2004).
67. Shen, H., Basar, T.: Optimal nonlinear pricing for a monopolistic network service provider with complete and incomplete information. IEEE J. Sel. Areas Commun. **25**(6), 1216–1223 (2007)
68. Shenker, S.: Fundamental design issues for the future Internet. IEEE J. Sel. Areas Commun. **13**(7), 1176–1188 (1995)
69. Simaan, M., Cruz, J.B.: On the Stackelberg strategy in nonzero-sum games. J. Optim. Theory Appl. **11**(5), 533–555 (1973)
70. Stidham, S.: Pricing and congestion management in a network with heterogeneous users. Technical report (2004).
71. Vakali, A., Pallis, G.: Content delivery networks: status and trends. IEEE Internet Comput. **7**(6), 68–74 (2003)
72. Wang, Q., Chiu, D.M., Lui, J.C.S.: ISP uplink pricing in a competitive market. In: Proceedings of International Conference on Telecommunications 2008, pp. 1–6. IEEE (2008).
73. Wang, X., Schulzrinne, H.: Pricing network resources for adaptive applications in a differentiated services network. In: Proceedings of INFOCOM 2001, vol. 2, pp. 943–952. IEEE (2001).
74. Wang, X., Schulzrinne, H.: An integrated resource negotiation, pricing, and QoS adaptation framework for multimedia applications. IEEE J. Sel. Areas Commun. **18**(12), 2514–2529 (2000)
75. Wang, N., Ho, K., Pavlou, G., Howarth, M.: An overview of routing optimization for Internet traffic engineering. IEEE Commun. Surv. Tut. **10**(1), 36–56 (2008)
76. Wroclawski, J.: The use of RSVP with IETF integrated services. Request for Comments (RFC) 2210 (1997).
77. Yaiche, H., Mazumdar, R.R., Rosenberg, C.: A game theoretic framework for bandwidth allocation and pricing in broadband networks. IEEE/ACM Trans. Netw. **8**(5), 667–678 (2000)
78. Yuksel, M., Kalyanaraman, S.: Pricing granularity for congestion-sensitive pricing. In: Proceedings of International Symposium on Computers and Communication 2003, pp. 169–174. IEEE (2003).

79. Yuksel, M., Kalyanaraman, S.: Distributed dynamic capacity contracting: a congestion pricing framework for diff-serv. Manag. Multimed. Internet **2496**, 198–210 (2002)
80. Yuksel, M., Kalyanaraman, S.: Distributed dynamic capacity contracting: an overlay congestion pricing framework. Comput. Commun. **26**(13), 1484–1503 (2003)

Chapter 2
Brief History of Pricing Model

In this chapter, we will introduce pricing factors based on cost analysis. Generally, there are three basic models in traditional best-effort networks which represent important factors in the pricing of QoS guaranteed network services.

Network service types can be divided into best-effort service and QoS mechanism related services. For different service types, pricing models should be suitable for charging [70], and mechanisms are required to ensure pricing implementation. To this aim, we use examples to introduce the matching between pricing and services and make a brief evaluation.

In microeconomics, price level depends on market environments or structures. In the network research area, besides considering the market, resource pricing is also affected by network service mechanisms and generally settled through interactions among various participants who optimize their utilities. On this basis, system optimization models and strategic optimization are introduced accordingly.

2.1 Basic Pricing Models

In this chapter, we will introduce pricing factors based on cost analysis. Generally, there are three basic models in traditional best-effort networks which represent important factors in the pricing of QoS guaranteed network services.

The first one is flat pricing. At the early stages of the Internet, users utilize a small quantity of network resources. Thus, ISPs aim to attract a large number of users and occupy the market. They generally adopt a unified price C (or flat fee [51]) to charge users based on access costs, which means that in a certain period of time, the users with the same access rate will be charged equally. Intuitively, as the simplest pricing model, it is easy to implement flat pricing and there is no need for complex statistical systems. Moreover, it can stimulate network usage since the fee is unchanged no matter how much data is transmitted. Thus, the charges can be predicted by users.

K. Xu et al., *Internet Resource Pricing Models*,
SpringerBriefs in Computer Science, DOI: 10.1007/978-1-4614-8409-7_2,
© The Author(s) 2014

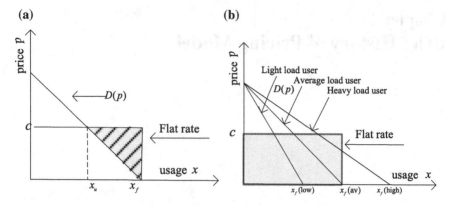

Fig. 2.1 **a** A customer will consume $D(p) = x_u$ units at a unit price of p, and x_u under a flat-rate charge. The *shaded area* represents the waste. **b** At a unit cost of c, the flat-rate charge is the *rectangle*. The *small triangle* is the value to the light user, and the *large triangle* is the value to the heavy user

However, drawbacks emerge as the traffic increases. As shown in Fig. 2.1a [23], suppose the unit cost of usage is c, the charge for user is p, and the demand curve is $D(p)$. And then we can find that:

(1) Users have no incentives to limit their usage, making network resources overused;
(2) Light users will compensate heavy users. If the flat fee C is charged based on the average usage amount, then $C = c \times x_f(av) = c \times D(0)$. And the costs for all users are shown as the rectangle area in Fig. 2.1b. Clearly, the costs of light-load users are higher than the gains, while the situation of the heavy-load users is on the contrary; and
(3) Resources are wasted to some extend. Estimated by the practical user utility, the usage over $D(c)$ will cause $\int_0^c [D(c) - D(0)]dp$ value loss, as the shade shown in Fig. 2.1a.

From the above discussion, we can infer that the flat pricing model is unable to help achieve optimized resource allocation. However, as one of the referential pricing factors, access charges can be used as a basic guarantee to recover fixed costs, which is still adopted by many ISPs.

As the usage and fixed costs have been distinguished and studied separately, usage-based pricing models can be discussed. Simply speaking, usage-based pricing means the charge P is related to the amount of resource usage. Currence et al. [17] believed that usage-based pricing can reflect actual use of network resources and is derived from traditional flat pricing.

Usage-based pricing was studied by a large number of researchers at early stages of the Internet [2, 17, 22, 37, 64, 73]. Generally, they used a supply-demand balance model in economics to describe the interactions between users and ISPs. Edell and Varaiya [23] showed in their experiments that users are highly sensitive to pricing, and thus usage-based pricing can enhance efficiency as well as guarantee fairness among users. Moveover, experiments in [22] illustrate that dynamic usage pricing can prevent congestion and improve the average network performance. However, other problems still need to be addressed, such as the privacy issues in processing audit and statistics [17] and the charge problem caused by user's non-expected traffic (e.g., Ads and Spam).

Practically, China Education and Research Network (CERNET) charges users full rate for international traffic [73]. Besides such direct traffic statistics, ISPs in general can use statistical sampling methods to estimate usage. For example, the 95th percentile pricing has been used as an industrial standard. And in this method, the peak flow within 5 % of the total time (36 h per month) is free of charge. Many ISPs adopt such peak flow rate based charging standards [17].

Recently, the overall user bandwidth demands have seen a dramatic increase. Consequently, increasingly differentiated usage patterns make the fairness problem even more serious, which indicates that it is more reasonable to charge heavy-load users according to usage [74]. However, in terms of P2P application providers who encourage users to participate in content sharing, such charging schemes will go contrary to their goals. So, more complicated interactions between P2P application providers and ISPs should be carefully studied. For example, He et al. [78] proposed a cooperative profit distribution method to avoid such conflict.

Intuitively, such a great number of concurrent network users may lead to higher system load. Researchers expect to constrain this negative external effect (named social costs [51]) through pricing. In other words, when the network is busy, the pricing is used to encourage users to avoid excessive resource usage in order to relieve or eliminate congestion [16, 27, 41–44, 51, 52, 54, 80]. The corresponding pricing is named congestion pricing.

Congestion pricing dynamically sets prices that can reflect approximate real-time network resource usage and represent current social costs. However, the measurement of such costs is not trivial, which requires detection of user's perceived value of marginal resources (like the shadow price in [41, 42]). And the cost cannot be directly calculated as the fixed one or measured based on usage. Thus it cannot be simply described by mathematical symbols.

In general network performance optimization articles, the congestion cost is described by delays in M/M/1 queuing system [3]. In Mason and Varian's smart market [51] pricing mechanism, an auction-based pricing method was proposed to measure and price such social costs. MacKie-Mason [52] further studied the advantages of smart markets using a generalized Vickrey auction mechanism [54] to allocate scarce resources. This kind of congestion pricing belongs to mechanism design (MD, [60]), which has been studied in incomplete information games. We leave out the details here.

The possibility of congestion can be reduced, since congestion pricing aims to implement network-aware pricing, which encourages shifting the traffic from peak time to non-peak time. However, as mentioned above, the implementation mechanism is always complex, and its effectiveness analyzed by Ykusel and Kalyanarama [80] is time-sensitive. They concluded that when the price interval is more than 40 times of RTT, the price can hardly affect congestion. In fact, time dependent usage-based pricing can also achieve a certain level of congestion control [22, 32], though it may not base on the analysis of social costs.

2.2 Pricing Mechanisms

Network service types can be divided into best-effort services and QoS mechanism related services. For different service types, pricing models should be suitable for charging [70] and mechanisms are required to ensure pricing implementation. In best-effort networks, user's fees are calculated by the access network, and as a result, pricing is done at the network edge, known as edge pricing [10, 70]. Typically, flat and usage pricing models are suitable for best-effort networks, whereas congestion pricing models are not. However, the weakness of edge pricing is that it cannot reflect network status. Thus, the effectiveness of pricing is limited. To solve this problem, several solutions are proposed. The main idea is that ISPs can negotiate with users in an access network based on expected congestion [70] or estimated traffic, instead of actual usage [11, 12]. Thus, it is easy to implement this pricing, which can prompt flexible interactions between ISPs and users. Unfortunately, due to the characters of distributed networks, although agreements exist, network-wide QoS guarantees or differentiation is hard to ensure.

However, through Paris Metro Pricing (PMP [61]) proposed by Odlyzko, we can attain prioritized services in a best-effort network. Basically, users want to enjoy better performance as much as possible with corresponding costs. As shown in Fig. 2.2 [24], the network is logically divided into channels with different transmission capacity C and corresponding price P. In principle, selecting channels at a higher price will get better service due to a smaller number of competitors. Meanwhile, since network providers divide users into different categories through charging, differentiated services are naturally achieved to some extent. However, PMP is only applicable to monopolistic networks. So, if the model is extended to a complex network environment, the pricing and resource sharing should be further studied.

Fig. 2.2 The PMP pricing

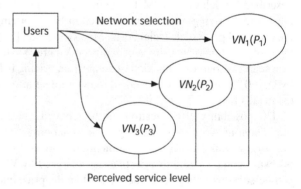

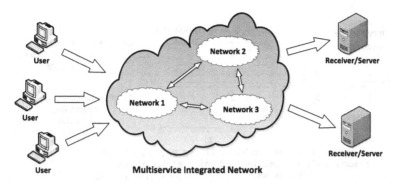

Fig. 2.3 Service class division based on QoS requirements

First introduced by Cocchi et al. [13, 14], simple priority-based service revealed the relationship between QoS differentiation and resource usage efficiency. To provide priority-based services, a reasonable way is to distinguish traffic by applications. And the simplest approach is to set priority levels and use Type of Service (ToS) fields in IP packets. Such a model is more realistic and implementable though QoS may not be guaranteed. However, since packet transmission for priority-based service depends on cooperation along the whole network path, coordination among ISPs is required.

Services can be divided into several classes according to their QoS requirements [19]. Therefore, those with higher QoS requirements will be given higher priority and charged a higher price. However, since the service price is pre-set here, when idle resources exist, users will still pay more for prioritized services without QoS guarantees.

Similarly, Donnell and Sethu [62] also suggested the priorities or service classes for data packets should be set by end user systems. Routers allocate them to different queues to ensure various service priorities. Gupta et al. [30] proposed a dynamic priority-based pricing mechanism and designed a real-time external price calculation method based on the congestion degree in a multi-class service environment. Their simulations show that dynamic pricing can significantly improve network performance and increase revenue. Furthermore, Gupta et al. [31] studied how to set an appropriate price to prevent users from distributing traffic into non-matching service classes, as shown in Fig. 2.3.

Priority-based service pricing can achieve average performance differentiation if the price and traffic are relatively stable during a long period of time. However, in a short term, high-priority service is more likely to experience more packet loss, longer delays, much more serious congestion, etc. To solve this problem, a proportionally differentiated service model which provides a relatively dynamic bandwidth division scheme was studied in [19, 20]. The main idea is that, as an extension of best-effort service type, the model will not strictly set bandwidth for each service class. Instead, it will use proportional performance guarantees to achieve a predictable and controllable QoS distinction (based on well-designed packet scheduling and packet

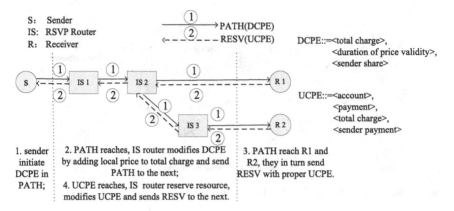

Fig. 2.4 Example of pricing session based on RSVP

discarding mechanisms). Thus, the corresponding proportional pricing is more applicable to such services.

In best-effort networks and simple priority-based service networks, QoS is not guaranteed. Accordingly, pricing usually depends on the actual cost or the resource usage. In contrast, this section will describe an Integrated Service (IntServ [77]) mechanism, which achieves QoS guarantee based on the resource reservation.

IntServ is a single-flow based architecture that can provide an end-to-end QoS guarantee. It uses end-to-end Resource Reservation Protocol (RSVP [82]) to reserve resources for each flow. Thus, the mechanism needs all routers to process per-flow signaling messages, maintain resource reservation status, and perform flow-based classification and scheduling. Specifically, routers first convert IP packets to traffic flows and then establish/dismantle resource reservation status for each flow according to whether existing resources meet the incoming flow's QoS requirements. If so, they implement QoS routing, corresponding scheduling and other controls to ensure the required QoS based on packet status.

Karsten et al. [40] studied a pricing mechanism applicable to RSVP. The main idea is to add price related information to regular RSVP messages so as to reserve resources and adjust the price. Specifically, the authors added Downstream Charging Policy Element (DCPE) to PATH messages and Upstream Charging Policy Element (UCPE) to RESV messages, where PATH and RESV are both regular RSVP messages. The mechanism works in the way shown in Fig. 2.4, from which we find that this pricing mechanism has more flexibility in sharing costs between senders and receivers. Therefore, it can support pricing for many applications including one- and two-side payments.

Similarly, Clark [11] proposed a zone-based charging or cost sharing model. Fankhauser et al. [26] proposed an RSVP-based accounting and charging protocol which is applicable to IntServ. They have proved that it can support local pricing models well through two pricing models: an auction-based pricing model (adding bid field of the RESV message), and a congestion sensitive usage-based pricing model.

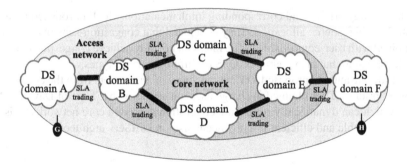

Fig. 2.5 Example of ISP networks at access and core levels

However, it needs to assume that the network performs static routing which will not be affected by the price and each pricing node in the network prices synchronously.

In fact, flow-based resource reservation is very complex and thus hard to achieve. Therefore, the realization of IntServ with QoS guarantees is not common and the corresponding pricing models are still under research.

As RSVP-based IntServ architecture has higher complexity and less scalability, Differentiated Services (DiffServ [8]) architecture is then proposed by IETF.

In DiffServ architecture, a complex flow control mechanism is realized at boundary nodes of the network and the process of inward nodes is simplified. Specifically, the boundary nodes conduct flow classification, shaping and aggregation, resulting in several flow aggregations first. The aggregation information is stored in a DS (Differentiated Service) field of IP packets called Differentiated Service Code Point (DSCP). And then, the internal nodes schedule and forward IP packets based on DSCP. As a hierarchical service structure, each DS region adopts an SLA (Service Level Agreement) and TCA (Traffic Conditioning Agreement) to coordinate and thus to provide cross-regional services. SLA clearly describes the supported service level and the allowed traffic volume at each service level, and TCA is used in detailed QoS negotiations.

Pricing is usually based on SLA in the DiffServ architecture. SLA can be a static or dynamic contract. In static SLA, regular consultations are needed, while in dynamic SLA, users need the signaling protocol (e.g., RSVP) to help request service dynamically. Fankhauser and Plattner [25] proposed an implementation profile to describe resource transactions in networks where a bandwidth broker acts as an SLA trader or negotiator.

Furthermore, Semret et al. established a double-layer DiffServ-based market model which considers users as bandwidth brokers and sellers within one market, as shown in Fig. 2.5 [66]. They concluded that driven by a dynamic market, bandwidth division among various service classes will finally be stable. Similarly, Wang and Schulzrinne proposed a framework named Resource Negotiation and Pricing (RNAP) [76]. They pointed out that pricing for reserved resources should be conducted differently on two levels. In [75], Wang and Schulzrinne built an optimization model to study pricing and corresponding implementation which introduced access control to aid resource allocation. They concluded that congestion-sensitive pricing combined with user-controllable traffic rates can not only achieve congestion control to a large extent, but also guarantee QoS of different service types.

In [48], the authors proposed a pricing mechanism that differs from the core/edge network pricing. They claimed to charge users on access with a Time of Day (TOD) price which can dynamically reflect the congestion degree in core networks. It is a flexible, scalable and efficient pricing mechanism in DiffServ architectures.

2.3 Pricing Methods

In this section, we will introduce two major network pricing methods that determine appropriate price levels: (1) system optimization models which are mainly based on a network utility maximization (NUM [9, 42]) framework; and (2) strategic optimization models, i.e., considering strategic behaviors of the others when setting prices or making other decisions [3, 4].

2.3.1 Pricing Based on NUM

From an economic point of view, an efficient market that refers to the total social welfare (i.e., the total surplus of service providers and users) is maximized [28]. Under different market environments, different conclusions can be drawn. We mainly introduce system utility (social surplus/welfare) optimization oriented pricing methods for a single network (unaffected by other providers) based on the optimization theory.

Kelly [42] proposed the concept of Network Utility Maximization (NUM) which is the initial work of Internet system optimization. The main object is to find the price that can make the total resource demand and supply in equilibrium. NUM framework can be described with three optimization problems which are shown as follow:

$$
\begin{aligned}
&A : \text{SYSTEM}[U, H, A, C] : \\
&\text{maximize} \qquad \sum_s U_s(x_s) \\
&\text{subject to} \qquad Hy = x, \ Ay \leq C \\
&\text{over} \qquad x, y \geq 0
\end{aligned}
\tag{2.1}
$$

where x_s denotes the traffic rate and U_s the value or utility of the traffic to user s. Service provider's cost is ignored. Then, the constraints are:

(1) $Hy = x$, where $H_{s \times r}$ denotes the source-destination pair $i \in \{1, 2, \ldots, s\}$ served by path $j \in \{1, 2, \ldots, r\}$, and vector $y = \{y_1, y_2, \ldots, y_r\}^T$ denotes the resources distributed to all source-destination pairs on each feasible path. This constraint means the whole distributed resources are equal to x_s for any user;

(2) $Ay \leq C$, where A is a 0-1 matrix telling whether the distributed resource is on the link, and the constraint means the sum of all distributed resources cannot exceed link capacity C;

(3) $x, y \geq 0$. Since user utility is unknown to the system, it is difficult to solve (A). In NUM, Kelly shows that the solutions of (A) are equal to that of two sub-optimization problems: user optimization (B) and network optimization (C).

$$B : \mathrm{USER}_s[U_s; \lambda_s] :$$
$$\text{maximize} \qquad U_s(m_s/\lambda_s) - m_s \qquad (2.2)$$
$$\text{over} \qquad m_s \geq 0$$

where λ_s denotes the price of per unit traffic charged to user s. Here, a user optimizes his surplus $U_s(m_s/\lambda_s) - m_s$ by deciding how much to pay m_s (which can be indirectly inferred by $x_s = m_s/\lambda_s$). For a network, it allocates network bandwidth to different flows according to user's feedbacks and some fairness standards shown as follow:

$$C : \mathrm{NETWORK}[H, A, C; m] :$$
$$\text{maximize} \qquad \sum_s m_s \log x_s$$
$$\text{subject to} \qquad Hy = x, Ay \leq C \qquad (2.3)$$
$$\text{over} \qquad x, y \geq 0$$

where H, A and C denote the network status with the same meaning in (2.1). Given $(m_1, m_2, \ldots, m_s)$, it tries to distribute bandwidths by solving (C) which seems to be based on weighted proportional fairness. Kelly pointed out that if $\forall s$, $U_s(\cdot)$ is a concave function, then this convex optimization problem has a unique optimal solution $x^* = (x_1^*, x_2^*, \ldots, x_s^*)$. $\lambda^* = (\lambda_1^*, \lambda_2^*, \ldots, \lambda_s^*)$, and $m^* = (m_1^*, m_2^*, \ldots, m_s^*)$, $m_s^* = \lambda_s^* x_s^*$ are validate for every $s \in S$. Then, the three optimization problems are all solved by their consistent solutions. The vector x^* is the unique optimal allocating rate and λ^* is the current optimal resource price vector.

System optimization problem (A) can also be decomposed into other types of sub-optimal problems. As its essence will not change, we just skip it here.

We have discussed the system model for elastic flows, where user utility are always described as concave functions. However, as a matter of fact, such willingness will vary with different types of applications. For example, for video and voice applications, if the transmission rate is less than a certain value, user's experience will decline sharply (as shown in Fig. 2.6 [46]). This indicates that the S-type utility function should be used to model user utility, and thus the convex optimization framework of NUM will no longer work. The resulting system can be seen as a hybrid service system, which includes inelastic flows. Therefore, the pricing and resource allocation problem becomes a difficult non-convex optimization problem [9, 46, 72].

To achieve the optimal system resource usage when heterogeneous flows coexist, Jang-Won et al. [46] first designed an incentive mechanism to inspire user's transmission cancellations. Such user behavior is called "self-regulation" which is similar

Fig. 2.6 Hybrid service system with various utility types

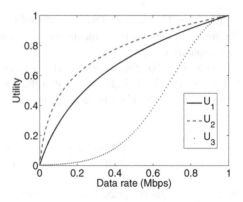

to access control of end systems. Mathematically, as the problem is non-convex and the duality gap may exist, the solution may not converge to optimality. Thus, an asymptotical optimal resource allocation algorithm is further designed by them.

Unlike the above approximate optimal solution, Chiang et al. [9] and Hande et al. [35, 36] studied rate allocation optimization framework for inelastic flows and presented sufficient and necessary conditions for the convergence to the global optimum of the proposed distributed rate allocation algorithm.

In fact, the NUM framework has also been applied to the edge pricing model. Currently, the sender (supplier) and the receiver (demander) may have different utilities to commute between them. So, ISPs need to set a supply-demand balanced price to maximize their revenue. Hande et al. [33] extended the NUM framework by adding content providers (CPs) to the system model. They concluded that regardless of network marketing environments (competition or monopoly), the overall system revenue and the utility of CPs will increase on the condition that CPs are charged for user compensation. The network neutrality issue was also discussed (NN [5]). Generally speaking, ISPs should not charge CPs differently according to content types.

2.3.2 Pricing Based on Game Theory

In real networks, there are three types of relationships: ISP-ISP, ISP-users, and user-user. Based on whether a binding agreement among them can be formed, games can be divided into non-cooperative games [57, 59] and cooperative games [7, 58, 65].

Considering non-cooperative games in network resource pricing and allocation, three levels of such interactions can be identified:

1. Competition among ISPs in the network market;
2. Leader-follower game between the ISP and users;
3. Resource competition among users.

Research on Multi-ISP interaction is facing great challenges. Therefore, comprehensive research results are still lacking today. In this section, we will mainly introduce non-cooperative game models for (2) and (3). For modeling relationships (2), in a monopoly network market, a single leader-follower game model (such as Stackelberg [3, 4, 67–69, 71]) is always applied. According to user utility known by ISPs, such work can be divided into two categories: pricing with complete or incomplete information. For modeling relationships (3), an N-person non-cooperative game is always employed. Here each person's behavior will affect the utilities of others.

Generally, in the leader-follower network resource pricing model, a leader (ISP) sets the price strategically, and the followers (users) act as price takers who decide how much resource to buy mostly based on the given price. The point here is the stable state (i.e., Nash Equilibrium [59]) where none of the participators wants to deviate from their behaviors when the strategies of others are known.

Specifically, in [3], they built two games at different levels: a non-cooperative game related to resource competition among users and a Stackelberg game where an ISP maximizes benefits within resource constraints based on the prediction of user reflection. In the first game, each user maximizes his/her goal described by the following Eq. 2.4 to decide rate:

$$F_i(x_i, x_{-i}; p) = w_i \log(1 + x_i) - \frac{1}{nc - \sum_j x_j} - px_i. \qquad (2.4)$$

where x_i is user's transmitting rate, nc is link capacity, $w_i \log(1 + x_i)$ is user utility function, $\frac{1}{nc - \sum_j x_j}$ represents congestion cost (i.e., queuing delay computed by using M/M/1 queuing model), and p is the unit price charged by ISPs. Then, they prove that user's non-cooperative game has Nash Equilibrium (NE). That is to say, for any user i, the solution x_i^* holds:

$$\underset{0 \leq x \leq nc - x_{-i}^*}{\text{maximize}} \; F_i(x_i, x_{-i}^*; p) = F_i(x_i^*, x_{-i}^*; p) \qquad (2.5)$$

which means that the decision made is the optimal one corresponding to all optimal decision of others.

In the second game, authors assumed that ISPs aim to maximize the benefits by solving Eq. 2.6, and thus to obtain the unit resource price p.

$$\underset{p \geq 0}{\text{maximize}} \; L(p; \overline{x}^*(p)), \; L(p; \overline{x}) := p \cdot \overline{x} \qquad (2.6)$$

where $\overline{x}^*(p) := \sum_i x_j^*(p)$ represents the total rates of all individuals in NE of a non-cooperative game. According to Eq. 2.4, since adding up all utility functions of users would not change the NE point, they derived a user equivalent optimization problem from Eq. 2.7:

$$F(x_1, \ldots, x_n; p) = \sum_{j=1}^{n} w_j \log(1 + x_j) - \frac{1}{nc - \overline{x}} - p\overline{x} \qquad (2.7)$$

where all utilities are added together. Then, by solving this convex optimization problem, a unique optimal solution $\overline{x}^*(p)$ can be obtained (notice that the solution is a function of price p). Finally, the above solution is deduced to Eq. 2.6 as a single-variable optimization problem. Solving it directly can obtain the optimal price p^*. The authors also gave an extended discussion in the case of multi-link afterwards [4].

Similar to the non-cooperative game framework mentioned above, Shen and Basar [69] extended the model to study non-linear optimal pricing in the cases of complete and incomplete information of user utility.

However, in a multi-ISP market, ISPs compete for users, and their prices are affected by other ISPs. Thus the model will become complex. Acemoglu and Ozdaglar [1] claimed that unlike the monopoly case where system efficiency can be improved and the social optimal is achieved at the equilibrium, in the multi-ISP competition game [28], the pure strategy NE may not exist (depending on the cost function).

Historically, the well-studied cooperation game models in network resource pricing are Nash Bargaining Game [58] and Shapley value [65] models. The former emphasizes Pareto optimal property and a certain sense of fairness; while the latter has well-formulated marginal contribution concept and the corresponding calculation methods. In recent years, as a new trend, such cooperative game models are studied and gradually applied to the modeling of network resource pricing [7, 49, 50, 55, 79].

Generally, NBS satisfies all the following four axioms [6, 58, 79]:

(1) Invariant to equivalent utility representations; (2) Pareto optimality; (3) Independence of irrelevant alternatives; and (4) Symmetry.

Therefore, it is usually applied to incentive efficient and fair cooperations where efficiency can be improved.

In [7], Cao et al. assumed that all network users have the same behavior characteristics and preferences. Thus, they simplified the pricing problem as a game between a single user and an ISP where each conducts optimization respectively. Then, they concluded that Nash bargaining can make the system operate at the Pareto efficient point with guaranteed fairness compared with the results in leader-follower games. Furthermore, in [79], the authors studied distributed network resource pricing and allocation, which we briefly introduce as follows:

First, ISPs face a centralized fair resource allocation problem which is formulated in accordance with the concept of Nash bargaining. It is shown as the following constrained convex optimization problem:

$$\text{maximize } \prod_{i=1}^{N} (x_i - MR_i)$$
$$\text{subject to } x_i \geq MR_i, \tag{2.8}$$
$$x_i \leq PR_i,$$
$$(Ax)_l \leq (C)_l.$$

where x_i is the resource (rate) assigned to user i, and MR_i and PR_i are the minimum and peak rate requirements of user i. However, such a central solution always brings in a lot of network communication burdens. Therefore, the authors proposed a distributed model where users optimize their own utilities with an added penalty $\alpha_i x_i$, and the aggregated rate is expected to ensure that the system can operate at the Pareto optimal point. Thus, for each user, Eq. 2.9 is optimized for rate selection:

$$\underset{x_i}{\text{maximize }} \ln(x_i - MR_i) - \alpha_i x_i$$
$$\text{subject to } x_i \geq MR_i, \tag{2.9}$$
$$x_i \leq PR_i.$$

Similar to the leader-follower game in Sect. 4.2.1, the network here needs to solve the rate allocation problem which can maximize its revenue. Besides, the revenue is calculated by the sum of penalties, as shown in Eq. 2.10. The constraints are the same in Eq. 2.8.

$$\text{maximize } \sum_{i=1}^{N} \alpha_i x_i \tag{2.10}$$

Obviously, the distributed method can maximize user utility as well as network revenue, which is similar to the results of the NUM-based model. However, the key difference is that their system objectives are different: one is to maximize social welfare, while the other is to fairly distribute resources.

The second cooperation game model is Shapley model. Shapley value emphasizes revenue distribution based on weighted marginal contribution of each entity in a group. As an axiomatic method [65, 50, 49], basically, it satisfies properties including efficiency, symmetry, fairness and dummy.

Proposed by L. S. Shapley in 1953, the Shapley value φ provides a unique payoff allocation satisfying some fairness criteria, which is defined as

$$\varphi_i = \frac{1}{N!} \sum_{\pi \in \Pi} \Delta_i(v, S(\pi, i)) \quad \forall i \in N \tag{2.11}$$

where Π is the set of all $N!$ permutations of N, and $S(\pi, i)$ is the set of players preceding i in the permutation π. Thus, the Shapley value of each player can be explained as the expected marginal contribution $\Delta_i(v, S(\pi, i))$.

Several studies emerge with respect to applying the theory to real network pricing or profit sharing [49, 50]. However, generally, an obvious drawback is its computational complexity (i.e., N participants needs 2^N scale of computations). Besides, it requires a centralized allocation process which will also make it less scalable.

2.4 Summary

In this chapter, we first describe three basic pricing models. In a flat pricing model, the fee is generally constant in a long period of time and used to recover the fixed cost. And usage-based fee is charged accordingly to recover the usage cost. As a dynamic pricing where prices are dynamically adjusted, congestion pricing is proposed for measuring and charging.

In fact, these three pricing models are not orthogonal, which means although they reflect different pricing factors, their functions can overlap to some extent. For example, Altmann and Chu [2] proposed a hybrid pricing model that combined flat and usage pricing. In this model, users can enjoy basic services at a basic flat rate, while their higher bandwidth demands will be charged by usage. They show that such a pricing model can improve network performance and increase revenue for ISPs.

Then, based on two types of network services considering Qos and not considering Qos respectively, we introduce two kinds of pricing mechanisms.

For best-effort services, it is believed that if edge pricing uses expected congestion information, it can achieve a certain degree of congestion control. Also, one can distinguish access bandwidths to provide certain prioritized services. But neither of them can ensure resource efficiency or gurantee QoS. It is more complex for QoS-based pricing. And the corresponding pricing process can be more difficult with higher complexity, especially for IntServ pricing, because QoS is guaranteed based on per-flow resource reservation. However, for DiffServ, QoS is guaranteed based on aggregated flows. So DiffServ improves the efficiency at a lower level of complexity compared with IntServ. Indeed, the combination of IntServ (in edge networks) and DiffServ (in core networks) to provide differentiated services can enjoy the benefits of low complexity and high efficiency with a certain degree of QoS.

At last, we classify and summarize typical pricing methods of network resources based on two main research paths: (1) The system optimization model is mainly based on the NUM framework. Considering network traffic characteristics, it can be divided into an optimization model for an elastic flow system and an optimization model for a hybrid system where inelastic and elastic flows coexist; and (2) Strategic optimization models are based on two major branches of the game theory non-cooperative games and cooperative games. We discuss two game theories here. The former proposes a cooperative game based profit maximization and distribution method in P2P markets, and the latter gives a non-cooperative game based pricing mechanism in the mobile markets of multi-interface users.

We briefly describe the relationship between the system optimization model and the non-cooperative game model. For the former, ISP dynamically controls the system through the pricing mechanism to help reach an optimal equilibrium. For the latter, the analysis of strategic behaviors of all participants based on non-cooperative game theory can help determine whether the system has NE and evaluate its efficiency.

References

1. Acemoglu, D., Ozdaglar, A.: Competition and efficiency in congested markets. Math. Oper. Res. **32**(1), 1–31 (2007)
2. Altmann, J., Chu, K.: How to charge for network service-flat-rate or usage-based? Comput. Netw. **36**(5), 519–531 (2001)
3. Basar, T., Srikant, R.: Revenue-maximizing pricing and capacity expansion in a many-users regime. In: Proceedings of INFOCOM 2002, vol. 1, pp. 294–301. IEEE (2002)
4. Basar, T., Srikant, R.: A Stackelberg network game with a large number of followers. J. Optimization Theory Appl. **115**(3), 479–490 (2002)
5. Beverly, R., Bauer, S., Berger, A.: The Internet is not a big truck: toward quantifying network neutrality. Passive Act. Netw. Measur. **4427**, 135–144 (2007)
6. Cao, X.: Preference functions and bargaining solutions. In: Proceedings of IEEE Conference on Decision and Control 1982, vol. 21, pp. 164–171. IEEE (1982)
7. Cao, X.R., Shen, H.X., Milito, R., Wirth, P.: Internet pricing with a game theoretical approach: concepts and examples. IEEE/ACM Trans. Netw. **10**(2), 208–216 (2002)
8. Carlson, M., Weiss, W., Blake, S., Wang, Z., Black, D., Davies, E.: An architecture for differentiated services. Request for Comments (RFC) **2475** (1998)
9. Chiang, M., Zhang, S., Hande, P.: Distributed rate allocation for inelastic flows: optimization frameworks, optimality conditions, and optimal algorithms. In: Proceedings of INFOCOM 2005, vol. 4, pp. 2679–2690. IEEE (2005)
10. Clark, D.: A model for cost allocation and pricing in the Internet. J. Electron. Publ. **1** (1995)
11. Clark, D.: Combining sender and receiver payments in the Internet. In: Interconnection and the Internet: Selected Papers From the 1996 Telecommunications Policy Research Conference, p. 95. Lawrence Erlbaum (1997)
12. Clark, D.: Internet cost allocation and pricing. In: McKnight, L., Bailey, J. (eds.) Internet Economics, pp. 215–252. MIT Press, Cambridge (1997)
13. Cocchi, R., Estrin, D., Shenker, S., Zhang, L.: A study of priority pricing in multiple service class networks. In: ACM SIGCOMM Computer Communication Review, vol. 21, pp. 123–130. ACM (1991)
14. Cocchi, R., Shenker, S., Estrin, D., Zhang, L.: Pricing in computer networks: motivation, formulation, and example. IEEE/ACM Trans. Netw. **1**(6), 614–627 (1993)
15. Courcoubetis, C., Stamoulis, G.D., Manolakis, C., Kelly, F.P.: An intelligent agent for optimizing QoS-for-money in priced ABR connections. http://citeseerx.ist.psu.edu/viewdoc/download?doi=10.1.1.22.291&rep=rep1&type=pdf (1998)
16. Crowcroft, J., Oechslin, P.: Differentiated end-to-end Internet services using a weighted proportional fair sharing TCP. ACM SIGCOMM Comput. Commun. Rev. **28**(3), 53–69 (1998)
17. Currence, M., Kurzon, A., Smud, D., Trłas, L.: A causal analysis of usage-based billing on IP networks. Tech. rep. (2000)
18. Dahshan, M.H., Verma, P.K.: Resource based pricing framework for integrated services networks. J. Netw. **2**(3), 36–45 (2007)
19. Dovrolis, C., Ramanathan, P.: A case for relative differentiated services and the proportional differentiation model. IEEE Netw. **13**(5), 26–34 (1999)

20. Dovrolis, C., Stiliadis, D., Ramanathan, P.: Proportional differentiated services: delay differentiation and packet scheduling. IEEE/ACM Trans. Netw. **10**(1), 12–26 (2002)
21. Dube, P., Borkar, V.S., Manjunath, D.: Differential join prices for parallel queues: social optimality, dynamic pricing algorithms and application to internet pricing. In: Proceedings of INFOCOM 2002, vol. 1, pp. 276–283. IEEE (2002)
22. Edell, R.J., McKeown, N., Varaiya, P.P.: Billing users and pricing for TCP. IEEE J. Sel. Areas Commun. **13**(7), 1162–1175 (1995)
23. Edell, R., Varaiya, P.: Providing internet access: what we learn from index. IEEE Netw. **13**(5), 18–25 (1999)
24. Falkner, M., Devetsikiotis, M., Lambadaris, I.: An overview of pricing concepts for broadband IP networks. IEEE Commun. Surv. **3**, 2–13 (2000)
25. Fankhauser, G., Plattner, B.: Diffserv bandwidth brokers as mini-markets. In: Proceedings of the Workshop on Internet Service Quality Economics, MIT, US (1999)
26. Fankhauser, G., Stiller, B., Vogtli, C., Plattner, B.: Reservation-based charging in an integrated services network. In: Proceedings of the 4th INFORMS Telecommunications Conference, vol. 302, pp. 305–309. (1998)
27. Gibbens, R.J., Kelly, F.P.: Resource pricing and the evolution of congestion control. Automatica **35**(12), 1969–1985 (1999)
28. Gregory, M.N.: Principles of microeconomics. South-Western Pub (2011)
29. Gupta, A., Stahl, D.O., Whinston, A.B.: A priority pricing approach to manage multi-service class networks in real-time. J. Electron. Publ. **1** (1995)
30. Gupta, A., Stahl, D.O., Whinston, A.B.: A priority pricing approach to manage multi-service class networks in real-time. J. Electron. Publishing **1**(1&2) (1995)
31. Gupta, A., Stahl, D.O., Whinston, A.B.: An economic approach to networked computing with priority classes. J. Organ. Comput. Electron. Comm. **6**(1), 71–95 (1996)
32. Ha, S., Sen, S., Joe-Wong, C., Im, Y., Chiang, M.: TUBE: Time-Dependent Pricing for Mobile Data. In: Proceedings of SIGCOMM 2012, pp. 247–258. ACM (2012)
33. Hande, P., Chiang, M., Calderbank, R., Rangan, S.: Network pricing and rate allocation with content provider participation. In: Proceedings of INFOCOM 2009, pp. 990–998. IEEE (2009)
34. Hande, P., Chiang, M., Calderbank, R., Zhang, J.: Pricing under constraints in access networks: Revenue maximization and congestion management. In: Proceedings of INFOCOM 2010, pp. 1–9. IEEE (2010)
35. Hande, P., Shengyu, Z., Mung, C.: Distributed rate allocation for inelastic flows. IEEE/ACM Trans. Netw. **15**(6), 1240–1253 (2007)
36. Hande, P., Rangan, S., Chiang, M., Wu, X.: Distributed uplink power control for optimal sir assignment in cellular data networks. IEEE/ACM Trans. Netw. **16**(6), 1420–1433 (2008)
37. Honig, M.L., Steiglitz, K.: Usage-based pricing of packet data generated by a heterogeneous user population. In: Proceedings of INFOCOM 1995, pp. 867–874. IEEE (1995)
38. Johari, R.: Efficiency loss in market mechanisms for resource allocation. Ph.D. thesis, Massachusetts Institute of Technology (MIT) (2004)
39. Johari, R., Tsitsiklis, J.N.: Efficiency loss in a network resource allocation game. Math. Oper. Res. **29**(3), 407–435 (2004)
40. Karsten, M., Schmitt, J., Wolf, L., Steinmetz, R.: An embedded charging approach for RSVP. In: Proceedings of the International Workshop on Quality of Service 1998, pp. 91–100. IEEE (1998)
41. Kelly, F.: Charging and rate control for elastic traffic. Eur. Trans. Telecommun. **8**(1), 33–37 (1997)
42. Kelly, F.P., Maulloo, A.K., Tan, D.K.H.: Rate control for communication networks: shadow prices, proportional fairness and stability. J. Oper. Res. Soc. **49**(3), 237–252 (1998)
43. Keon, N.: A new pricing model for competitive telecommunications services using congestion discounts. INFORMS J. Comput. **17**(2), 248–262 (2005)
44. Kunniyur, S., Srikant, R.: End-to-end congestion control schemes: Utility functions, random losses and ecn marks. IEEE/ACM Trans. Netw. **11**(5), 689–702 (2003)

45. La, R.J., Anantharam, V.: Utility-based rate control in the internet for elastic traffic. IEEE/ACM Trans. Netw. **10**(2), 272–286 (2002)
46. Lee, J.W., Mazumdar, R.R., Shroff, N.B.: Non-convex optimization and rate control for multi-class services in the internet. IEEE/ACM Trans. Netw. **13**(4), 827–840 (2005)
47. Li, S., Huang, J., Li, S.Y.R.: Revenue maximization for communication networks with usage-based pricing. In: Proceedings of Global Telecommunications Conference 2009, pp. 1–6. IEEE (2009)
48. Li, T., Iraqi, Y., Boutaba, R.: Pricing and admission control for QoS-enabled Internet. Comput. Netw. **46**(1), 87–110 (2004)
49. Ma, R.T.B., Chiu, D., Lui, J., Misra, V., Rubenstein, D.: Interconnecting eyeballs to content: a Shapley value perspective on ISP peering and settlement. In: Proceedings of the 3rd International Workshop on Economics of Networked Systems, pp. 61–66. ACM (2008)
50. Ma, R.T.B., Chiu, D.M., Lui, J.C.S., Misra, V., Rubenstein, D.: On cooperative settlement between content, transit, and eyeball Internet service providers. IEEE/ACM Trans. Netw. **19**(3), 802–815 (2011)
51. MacKie-Mason, J.K., Varian, H.R.: Pricing the Internet. http://people.ischool.berkeley.edu/hal/Papers/UM/Pricing_the_Internet.pdf
52. MacKie-Mason, J.: A smart market for resource reservation in a multiple quality of service information network. Available at SSRN 975871. http://papers.ssrn.com/sol3/papers.cfm?abstract_id=975871 (1997)
53. MacKie-Mason, J.K., Varian, H.R.: Pricing congestible network resources. IEEE J. Sel. Areas Commun. **13**(7), 1141–1149 (1995)
54. Milgrom, P.: Putting auction theory to work. Cambridge University Press, Cambridge (2004)
55. Misra, V., Ioannidis, S., Chaintreau, A., Massouli, L.: Incentivizing peer-assisted services: a fluid shapley value approach. In: ACM SIGMETRICS Performance Evaluation Review, vol. 38, pp. 215–226. ACM (2010)
56. Mo, J., Walrand, J.: Fair end-to-end window-based congestion control. IEEE/ACM Trans. Netw. **8**(5), 556–567 (2000)
57. Myerson, R.B.: Game theory: analysis of conflict. Harvard University Press, Cambridge (1997)
58. Nash Jr, J.F.: The bargaining problem. Econometrica: J. Econ. Soc. **18**, 155–162 (1950)
59. Nash, J.: Non-cooperative games. Ann. Math. **54**(2), 286–295 (1951)
60. Nisan, N., Ronen, A.: Algorithmic mechanism design. In: Proceedings of the Thirty-first Annual ACM Symposium on Theory of Computing, pp. 129–140. ACM (1999)
61. Odlyzko, A.: Paris metro pricing for the Internet. In: Proceedings of the 1st ACM Conference on Electronic Commerce, vol. 3, pp. 140–147. (1999)
62. O'Donnell, A.J., Sethu, H.: Congestion control, differentiated services, and efficient capacity management through a novel pricing strategy. Comput. Commun. **26**(13), 1457–1469 (2003)
63. Ozdaglar, A., Srikant, R.: Incentives and pricing in communication networks. Algorithmic Game Theory, pp. 571–591. Cambridge University Press, Cambridge (2007)
64. Roberts, J.W.: Quality of service guarantees and charging in multi-service networks. IEICE Trans. Commun. **81**, 824–831 (1998)
65. Roth, A.E.: The Shapley value: essays in honor of Lloyd S. Cambridge University Press, Shapley (1988)
66. Semret, N., Liao, R.R.F., Campbell, A.T., Lazar, A.A.: Pricing, provisioning and peering: dynamic markets for differentiated Internet services and implications for network interconnections. IEEE J. Sel. Areas Commun. **18**(12), 2499–2513 (2000)
67. Shakkottai, S., Srikant, R., Ozdaglar, A., Acemoglu, D.: The price of simplicity. IEEE J. Sel. Areas Commun. **26**(7), 1269–1276 (2008)
68. Shen, H., Basar, T.: Differentiated internet pricing using a hierarchical network game model. In: Proceedings of the American Control Conference 2004, vol. 3, pp. 2322–2327. IEEE (2004)
69. Shen, H., Basar, T.: Optimal nonlinear pricing for a monopolistic network service provider with complete and incomplete information. IEEE J. Sel. Areas Commun. **25**(6), 1216–1223 (2007)

70. Shenker, S., Clark, D., Estrin, D., Herzog, S.: Pricing in computer networks: reshaping the research agenda. Telecommun. Policy **20**(3), 183–201 (1996)
71. Simaan, M., Cruz, J.B.: On the Stackelberg strategy in nonzero-sum games. J. Optim. Theory Appl. **11**(5), 533–555 (1973)
72. Stidham, S.: Pricing and congestion management in a network with heterogeneous users. IEEE Trans. Autom. Control **49**(6), 976–981 (2004)
73. URL http://www.cernet.edu.cn/20010912/3001298.shtml
74. Wang, Q., Chiu, D.M., Lui, J.C.S.: ISP uplink pricing in a competitive market. In: Proceedings of International Conference on Telecommunications 2008, pp. 1–6. IEEE (2008)
75. Wang, X., Schulzrinne, H.: Pricing network resources for adaptive applications in a differentiated services network. In: Proceedings of INFOCOM 2001, vol. 2, pp. 943–952. IEEE (2001)
76. Wang, X., Schulzrinne, H.: An integrated resource negotiation, pricing, and qos adaptation framework for multimedia applications. IEEE J. Selected Areas Comm **18**(12), 2514–2529 (2000)
77. Wroclawski, J.: The use of RSVP with IETF integrated services. Request for Comments (RFC) **2210** (1997)
78. Xu, K., Zhong, Y.F., He, H.: Can P2P technology benefit ISPs? A cooperative profit-distribution answer. http://arxiv.org/abs/1212.4915
79. Yaiche, H., Mazumdar, R.R., Rosenberg, C.: A game theoretic framework for bandwidth allocation and pricing in broadband networks. IEEE/ACM Trans. Netw. **8**(5), 667–678 (2000)
80. Yuksel, M., Kalyanaraman, S.: Pricing granularity for congestion-sensitive pricing. In: Proceedings of International Symposium on Computers and Communication 2003, pp. 169–174. IEEE (2003)
81. Yuksel, M., Kalyanaraman, S.: Distributed dynamic capacity contracting: an overlay congestion pricing framework. Comput. Commun. **26**(13), 1484–1503 (2003)
82. Zhang, L., Berson, S., Herzog, S., Jamin, S.: Resource reservation protocol (RSVP)-Version 1 functional specification. Request for Comments (RFC) **2205** (1997)

Chapter 3
Pricing and Management Related to P2P and Mobile Internet

With the rapid development of the Internet, the amount and types of network applications have increased significantly, and hence the traffic. On the other hand, consumer's demand for network resources is also becoming more and more intense. Although easy to implement, the traditional flat pricing makes the utilization rate of network resources drop significantly. Especially since the appearance of new network applications like P2P, the cost of network operation has soared, while the profit of ISPs has decreased notably. As a countermeasure, many ISPs, such as AT&T, Verizon and Comcast have abandoned the traditional flat pricing and adopted the traffic-based hierarchical pricing method instead [43], which has thus become a hot topic. At the same time, as the Internet market has become more mature, the competition between ISPs and content providers (CPs) have been brought into full play. Therefore, games appear frequently in resource pricing analyses.

In the previous two chapters, we summarized the basic pricing model and pricing mechanisms. In this chapter, we will briefly introduce P2P and the development of mobile Internet and analyze the challenge they brought to network management and pricing mechanism. In Chaps. 4 and 5, we will give two specific research instances. The first one focuses on the imbalance of profit distribution brought by P2P traffic through the cooperation between ISPs and CPs. While the later one discusses the problems in pricing games between ISPs in different kinds of markets.

3.1 P2P's Impact on Cooperation-Based Internet Pricing and Profit Distribution

As a new content distribution technology, P2P brings an innovation for network traffic models while satisfying the growing demand for high-bandwidth applications [42]. Such changes make the traditional resource pricing method and profit distribution mechanism of client/server (C/S) scheme encounter new challenges.

K. Xu et al., *Internet Resource Pricing Models*,
SpringerBriefs in Computer Science, DOI: 10.1007/978-1-4614-8409-7_3,
© The Author(s) 2014

Fig. 3.1 Comparison
between C/S and P2P

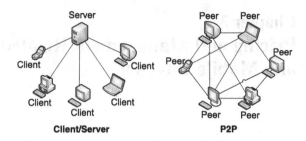

3.1.1 Development of P2P

As shown in Fig. 3.1, different from the traditional C/S scheme, P2P attracts more
and more applications with its distributed nature, good scalability and economy
(free riding). However, although P2P helps CPs (PCPs) save the cost substantially, it
has brought severe impacts on the profit and profit distribution of ISPs. So far, many
researchers have discussed P2P's influence on network participants, for example [57].
The key problem within is the potential influence in economy. We will introduce the
related work according to the problems below (e.g., ISPs take economic measures
to P2P) and make a distinction between them and the work in Chap. 4.

P2P technology is adopted to improve the content distribution. The effectiveness
has been verified in mercantile system [40, 41, 47] and the corresponding measure-
ments [18, 20, 27]. Related services are P2P file download, P2P streaming media,
P2P-VOD and P2P-IPTV. Related work [11, 20, 21, 46, 48, 58] also gives suggestions on system design to help CPs save more server bandwidth while using auxiliary
resource in content distribution so as to guarantee user experience.

Hei et al. [18] made a small-scale measurement on P2P-IPTV application in large-
scale practical system (PPLive [40]). By arranging PPLive crawler and package
detection software such as siners in the access points of campus networks and retail
broadband networks, they measured the PPLive traffic during the Spring Festival in
China by combining active and passive measurement technologies together and then
speculated the network structure of streaming architecture based on the data. The
authors found that P2P video streaming shall bring redundant data transmission to
the Internet. What's more, the successful deployment of P2P-IPTV and the client's
upload ability are closely related to each other. Consequently, there will be enormous
challenges on upload capabilities connected to an ISP.

Huang et al. are the first to introduce the measurement and analysis of the large-
scale on-demand video system. They utilized the data in Microsoft MSN system in
[21], and then analyzed the essential feature of MSN video-on-demand service in
C/S scheme. They pointed out peer-assisted content distribution will make it possible
to reduce the server bandwidth from 2.20 Gbps (obtained from the 95 percentile in
December 2006). (ISPs usually adopt the 95 percentile to charge the accessing CPs
or ISP clients [15]) to 79 Mbps. Thus, they can reduce the content distribution cost
substantially (ISPs or CDNs charge 0.1–1.0 cent to the video 200,440 kbps per

minute). In this paper, the authors concluded that if the video quality is increased by three times, the server can still save 48.5 % of the bandwidth.

By measuring methods, Huang et al. [20] analyzed the challenges and architecture design problems in guaranteeing the quality of service while saving server bandwidth in the actual mass P2P-VoD system (PPLive [40]). They then discussed a variety of optimization measures and the related system implementation issues. Similarly, in the process of content distribution, the servers have a backup complementary role of P2P. During the measurement in the actual P2P-VoD system, the authors found that the adoption of weighted copy strategy, proper transmission strategy and scheduling policy can further reduce server bandwidth in the system. In the measurement of a day, the authors concluded that the server's average load occupied about 8.3 % of the whole system. So we infer that in the actual system, P2P can help CPs (PCPs) save a lot of cost. Cheng et al. introduced the design, implementation and evaluation of the actual P2P-VoD system Gridcast arranged in CERNET. They focused on analyzing the optimization made by cache and copy of the replication strategy on the decrease of the server bandwidth. At last, they concluded that in the case of multiple video caching (MVC) and pre-copy, the average server bandwidth can reduce at least 51 %.

Compared to the cost reduced by P2P, we can infer negative effects of P2P traffic on local ISPs.

Taking MSN video as an example, Huang et al. [21] combined the recognition of P2P auxiliary traffic distribution, peer's IP address mapping in ISPs and the supposed economic relationships between ISPs together, and then studied the economic effect on ISPs made by peers during random or ISP-friendly neighbor selection (mainly from the traffic across the net settlement). Experiments show that under the random condition, most P2P traffic transfer across the network, which will bring ISPs huge cost. However, when decreasing ISP settlement across the net via limiting peer selection to the maximum degree, peer auxiliary can still reduce the server bandwidth to at least 50 %. However, the traffic provided by users would make the ISP cost unable to be compensated in the ISP flat charging model. In addition, although this work cannot explain PCP's economic influence on ISPs in quantification, it provide references on identifying application flow and estimating economic impact.

Different from the work above, Karagiannis et al. [23] analyzed P2P's influence on ISPs, PCPs and users by measuring the BitTorrent system [10]. Besides the same conclusion as the work above (peer auxiliary content distribution will benefit PCPs and users, but it will not increase the income of ISPs), the authors also estimated the effectiveness of locality-aware P2P implementation (the neighbor selection method peers accessing to content from the Internet). Through quantitative analysis, the authors believe this method can achieve the approximate cache effect. So we can infer that the outlet flow and the corresponding cost of ISPs can be reduced. But this work made the treatment for the mass upload network flow more difficult.

Many articles [1, 21, 34, 42, 45, 51, 52, 55] discussed the negative effects made by P2P and ISP's possible measures. For example, the negative resistance to P2P traffic [34], and the cooperation with PCPs on projects [1, 45, 55]. Some also introduced how to reduce the impact of ISPs from the perspective of PCPs, such as the ISP-friendly neighbor selection [21] and the CDN based neighbor selection [12].

Kaya et al. [24] analyzed the risk and feasibility in the above project including the ISP and PCP cooperation and the PCP auxiliary neighbor selection based on the third party information (CDN). They pointed out the potential abnormal result especially in the case of multiple ISPs. For instance, in the customer–supplier relationship, when providing P2P-related oracle information to select neighbors, adding selfish preferences will make ISP customers facing high costs and congestion. Aimed at these problems, the authors gave some suggestions on system improvement, however, the effectiveness still need further analysis and validation.

Actually, the primary cause of the negative effect of ISPs is P2P free-riding, which makes the cost of ISPs with heavy P2P traffic load unable to receive payoffs and even increase the ISP (usually access ISP) settlement cost across the network due to randomness, forming significant contrast with PCP's benefit. Inappropriate interest distribution makes ISP's interest damaged and thus the contradiction with P2P is sharpened.

3.1.2 Charging and Profit Division in P2P Networks

In this book, we try to solve the imbalance problem of interest distribution mentioned above by economic means. And the aims are as follow: on the one hand, we can promote ISPs to support P2P by providing rational economic payoffs; on the other hand, from the perspective of promoting the advantages of P2P technology, we can encourage PCP system to optimize system performance and help reduce the network transmission cost.

Different from this book, Wang et al. [52] analyzed P2P's influence on the peering relationship between ISPs. They abstracted networks as follow: as customers, two local ISPs buy bandwidth from the upper ISP (the provider) and then provide accesses to the users. The relationships between ISP-ISP and ISP-users are studied, as shown in Fig. 3.2. In short, this paper adopts multi-leader-follower game model for the interaction in ISP-ISP (user's pricing affects user market share of each other) and ISP-user (ISP flat pricing and network performance decide user's choice of ISPs). When users' network utility (decided by network performance and price) in the two ISPs are the same, the market reaches equilibrium. (The user can achieve the same effect no matter which ISP is chosen.) Through the instance, the authors analyzed the benefit of ISPs and the relationship between the variables at the network equilibrium and then gave some bandwidth suggestions on local ISPs and whether there are peer relationships between them. The authors concluded that even if there is a tremendous amount of P2P network traffic, the equivalence relationship between local ISPs is still good for the overall local ISPs. In other words, with the help of large ISPs' P2P users, smaller ISPs will get free-riding.

Based on a simple game model of dual ISP competition in the market, Wang et al. [51] analyzed the condition upload pricing instead of flat pricing in the ISP competition environment. The authors' opinion is that in peer-to-peer (P2P) content distribution mode, users make content distribution for CPs, which shall assume the

Fig. 3.2 Business model and game theoretic framework

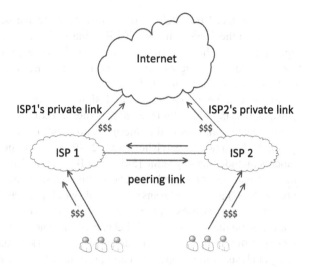

original CP fees for the Internet. However, due to the competitive market environment, users can choose from different ISPs. So pricing is not easy to implement. That is, ISPs should consider user loss in the market to pursue maximal benefit. Therefore it is hard to determine the appropriate price based on usage. In the guarantee of number of users, the authors gave the upload pricing in a single ISP to users based on the profit-neutral assumption. And then the influence on ISPs in the charge strategy is studied based on usage. The three potential results of the market (both flat pricing, both upload pricing, and both under dual charging modes) are also analyzed.

Actually, the work above all ignores the user behaviors in P2P systems. They usually assume that users will voluntarily provide necessary resources (such as storage and bandwidth) in the system. Another way is the mandatory implementation of user resource scheduling by P2P internal mechanism. However, in practice, as selfish network participants, users need to consider the optimization of their own benefit. At this time, appropriate client incentive mechanism is of great importance to the success of P2P systems. For example, the analysis in VoD systems by Ma et al. In [28], the Tit-for-Tat in BitTorrent, and better services provided to users willing to upload by Huang et al. [21], which all aim at promoting user participation and achieving system goals preferably. Expectations of incentive mechanism is simple and feasible in engineering, with less cost, better stability and performance.

Ma et al. [32] provided the evaluation framework of the incentive mechanism for P2P systems, which can be used to guide the actual design. Taking two incentive strategies based on historical information sharing as example, the authors carried out simulation analyses to their long-term stability and performance and then concluded that the proportional combination incentive mechanism of consumption and contribution factors will reach robust and scalable systems, while the image incentive mechanism would lead to less contribution of the user, and system crash eventually.

It is observed that PCP is often incentive mechanism design based on user's response to the service. In fact, if ISPs charge the user based on usage (such as the upload pricing), the user will consider not only the acquired service but also the cost for ISPs while deciding whether to contribute resources or not. Thus, PCP's incentive effect on users will be affected [42].

In this scheme, we can expect to design a cooperative and interest sharing mechanism satisfying both efficiency and fairness. Actually, due to the benefits of cooperative game solutions in the Internet, optimization model and protocol design has gradually become a research hot spot [22, 29, 31]. Commonly used cooperative game models include Shapley value [54], Nash Bargaining [36] etc. According to whether benefits can be transferred, the cooperative game is divided into non-transferable utility (NTU) games and transferable utility (TU) games.

This section focuses on the studies of fair benefit/profit division based on cooperative game theory. (Note: In NTU games, currency can be used to transfer benefits among different participants.) Currently, such work is mainly focused on the modeling analysis of the cooperation and profit division between ISPs and CPs.

Ma et al. [29, 31] proposed a profit division mechanism among ISPs through Shapley value [54] in cooperative game theory. Shapley value, as an axiomatic approach, is the unique solution that meets multiple properties (such as fairness and efficiency). Specifically, it divides the cost and benefit apportioned in accordance with the marginal contribution. In other words, the cost and benefits assigned to a player is the average of his/her marginal contribution to the union. Ma et al. defined the function worth of the union and the marginal contribution of the ISPs in the union. They found the profit division based on Shapley value can promote the ISPs to interconnect with each other and encourage the ISPs maximizing their own worth to adopt optimal routing strategy so as to achieve the profit maximization of the system. However, the realization of Shapley value is of high complexity because it needs a fair arbitration institution to assess the contribution of the players, which is very difficult in reality and incompatible with the existing settlement mechanism. Therefore, problems still exist in the application of Shapley value.

Misra et al. [35] proposed an incentive mechanism in the peer-assisted content delivery service based on the fluid Shapley value. They found that when the network owns a large number of users (peers), the assigned value of each participant will approach to limits, thus simplifying the computation of the Shapley value mechanism. Unlike [29, 31] which only consider problems from the perspective of ISPs, the participation of CPs and users are also taken into account. However, the difference is that their work is based on the assumption of the characteristics of each function, which is equivalent to giving the conditions of the satisfactory solution (such as nuclear [39]) (Note: a solution concept in cooperative game theory, which need to meet some properties). Comparing the gains of ISPs and CPs in cooperative situation and non-cooperative situation, Altman et al. [2] found that ISPs and CPs always gain a higher profit through cooperation than that in the non-cooperative situation. However, the main purpose is not to study the cooperation based profit division but the effectiveness of various factors on net neutrality when ISPs charge CPs volume-based rates. Thus, in this work, the ISPs and the CPs divide their revenue based on the simplest equal division.

Altman et al. [3] studied how should ISPs adjust the pricing of CPs through Nash Bargaining in the non-neutral network. Then, Altman et al. [4] studied this problem again in a more general scenario. However, the main purpose of both work is to discuss the effect of different bargaining power of the ISPs and the CPs on the charge amount.

We will propose our model in Chap. 4. Different from the above-mentioned studies, our model quantifies the revenue change of PCPs and ISPs caused by P2P and discusses the boundary condition of the ISPs taking actions to increase their own revenue. Then, we will analyze the revenue change after ISPs change their pricing strategy on users. Apparently, the revenue of PCP will be affected by ISP's pricing strategy. In addition, the differences between our model and the original methods are: compared with the work in [35], we consider the general models of ISPs and CPs' user demand and profits and give a general framework to analyze the effect of the peer-assisted content delivery on players' revenue, helping judge whether they would participate in the cooperation according to individual rationality; compared with the work in [2], the profit distribution model proposed in this book is based on NBS (Nash bargaining solution); compared with the works in [3, 4], the purpose of the model in this book is to promote the cooperation between ISPs and PCPs, and thus stimulating the continuous growth of P2P applications.

In Chap. 4, we will give the detailed description of the model. It quantifies the unbalanced profit distribution between PCPs and ISPs and analyzes the triggered strategic interactions. This model is used to predict the possible system states with non-cooperation, so as to prepare for the design of a fair profit distribution mechanism in the development of P2P applications.

3.2 Pricing Mechanisms in Mobile Internet

Mobile Internet is a major trend of network development. For example, the major ISPs in Chinese mobile Internet market include China Mobile, China Telecom and China Unicom. As of the end of December 2012, the number of Chinese netizens has reached 564 million and the number of mobile phone users 420 million, with an annual growth rate of 18.1 %. Various indicators of mobile network growth rate go beyond that of the traditional networks, and mobile phones grow fast in the areas of Weibo and E-commerce applications. By the end of December 2012, the size of the Weibo users in China has mounted to 309 million with an increase of 58.73 million compared with the end of 2011. The number of Weibo users accounts for 54.7 % of all the Internet users in China. The number of mobile Weibo users reaches 202 million, accounting for 65.6 % of all Weibo users. We can see that social network softwares such as Weibo has become an essential part of life for the majority of users. According to the statistics from China Internet Network Information Center, the distribution of the mobile social network user activity is shown in Fig. 3.3 [14].

Fig. 3.3 Mobile social network user activity in China

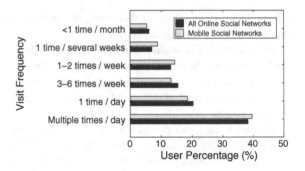

Mobile hosts using network services generally have multiple network interfaces to access the Internet through different ISPs [13]. However, interests lead to disputes. ISPs and Internet companies fight each other for the amount of users and profits. Meanwhile, users also have the right to choose different network services provided by different ISPs and Internet companies. Recently, WeChat, a typical OTT application in China, falls into a debate with mobile ISPs on the distribution of profits.

WeChat is a smart phone application launched by Tencent on January 21st 2011, which supports the transmission of voice SMS, videos, pictures, texts and group chats. This application is completely free except for the data traffic generated by messages, which results in a quick double of user size. As of January 23rd 2013, the number of WeChat users has been up to over 300 million. It is the rapid spread of WeChat that caused the recent debate on the pricing issue. In fact, the spearhead of the issue is not like the network rumors that Tencent will charge users, but that ISPs hope to charge Tencent on WeChat. Then what should China Mobile charge Tencent for? According to China Mobile, the excessive signals in the network will result in "signal storm". Besides, WeChat will continuously send signals to the base station to inform its online state and position, which is called the heartbeat state. Moreover, according to the interconnection system in China, China Mobile will always pay for the traffic between different ISPs in order to use the high profits of the mobile business to subside fixed network infrastructures. A part of the traffic generated by WeChat is actually paid by China Mobile. Meanwhile, with the free communication provided by WeChat, the business of China Mobile is also threatened. What might happen next is still hard to predict. But it can be seen that the core issue here is billing and profit distribution.

The complicated relationship between ISPs and CPs is evident. From an economic perspective, a dynamic game exists between ISPs and service users. In economics, the market is divided into a perfectly competitive market [38] and an imperfectly competitive market [17] according to the types of competition. Perfect competition is the ideal state of the market, which requires a sufficient number of sellers and buyers, becauseeach person alone cannot influence the market price. Imperfect competition is put forward by American economist J.M. Clark, who pointed out that perfect competition is an ideal state which does not exist in real life.

The network market we refer to does not meet the preconditions of a competitive competition market. First, the prices in the network market is impossible to stay constant regardless of the entry and exit of any ISP; Second, the services provided by different ISPs are impossible to be completely homogeneous, because R&D technologies may vary and thereby introduce different types of services; Finally, market participants cannot fully understand all the information, such as service cost and user income. In summary, network market is an imperfectly competitive market.

A certain degree of monopoly exists in an imperfectly competitive market. Market monopoly can be divided into monopolistic market [8], oligopolistic market [16] and joint monopoly market [37] according to organizational forms.

For network markets, a monopolistic market has only one ISP to provide services; an oligopolistic market has multiple ISPs to provide a variety of service combination, and users can select among multiple ISPs and service portfolios according to their needs; joint monololistic market has multiple ISPs to jointly control the network service market. Market monopoly will impede the competitive process and do harm to the society, and therefore are punished by many national laws [6]. From the perspective of game theory, joint monopolistic market is unstable. When an ISP adopts advanced production technology to reduce costs and adjust service portfolios to gain more users and profits, this form of monopoly will be broken. Once such a price coalition is broken, the market will turn into an oligopolistic market.

Many researchers studied the ISP service composition model. For example, three basic network resource pricing models of CDN and P2P networks, including flat pricing, usage-based pricing and congestion-based pricing, are surveyed in [19]. The survey also refers to hybrid pricing models, such as the one based on flat pricing and usage-based pricing [5]. In Chap. 5, we will introduce the service model instance that adopts such hybrid pricing model.

In addition, the researches on service pricing methods [5, 19, 25, 30, 33, 49, 50, 56, 59–61] depend on two theories: system optimization theory and game theory.

First, the goal of the system optimization theory is to achieve a network utility maximization [19, 25, 49, 61], which has been deeply studied.

For example, researchers in [49] studied the tiered pricing between upper tier ISPs and lower tier ISPs, as shown in Fig. 3.4. And the goal is to analyze whether the current tiered pricing is close to the optimal in the entire network transmission market and whether the optimal strategy exists. Tiered pricing improves both the profits of ISPs and the surplus of users. This paper proposed a user demand model and an ISP cost model. But it only solves the problem of ISPs, without the choice of users on the services.

Second, studies based on the cooperative game theory in economics are also in-depth and comprehensive.

Researchers proposed a network pricing philosophy based on cooperative game theory [56], pointing out that the bargaining game in QoS model is able to bring a better macro result, while the leader-follower game is not a Pareto optimal solution. There are many mature studies based on cooperative game theory. But economists found that cooperative relationships generally do not exist among participants in

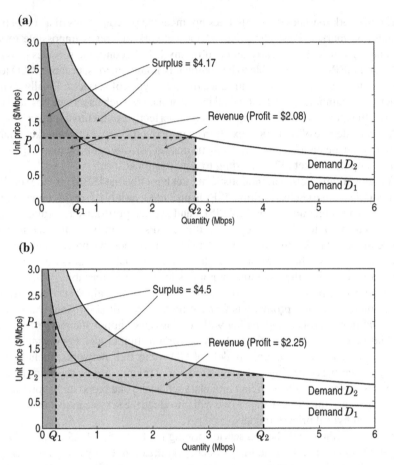

Fig. 3.4 Market efficiency loss due to coarse bundling [49] **a Blended-rate pricing**. IPS charges a single blended rate p_o. **b Tiered pricing**. ISP charges rates p_1 and p_2 for flows

reality. Few studies on the dynamic game between ISPs and users are based on non-cooperative game theory.

Offering various packages is a common way for ISPs to charge users. In fact, the package design is based on the mechanism design in game theory. For a single type of service, such as data service, ISPs usually adopt mechanism design to solve the information asymmetry problem and gain maximum profit through the design of optimal nonlinear pricing mechanisms [44]. In the mobile market, ISPs will provide various types of services, such as voice service, SMS service and data service. Many researchers have studied how providing a combination of services could bring potential profits for ISPs [7]. In April 2013, China Telecom, one of the largest mobile service providers in China, starts to provide the building block service compositions, allowing users to choose highly customized service compositions, as shown in Fig. 3.5.

Fig. 3.5 "Building block
package" by China Telecom

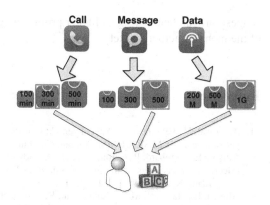

On the study of user utility in multi-interface mobile host, researchers [13] proposed a flow allocation method based on the active probing optimization of cost
and performance of the available paths by simplifying and modeling communications between the mobile host and the Internet. In terms of ISPs' cost computation,
Valancius et al. [48] referred that, in view of uncertainties number of basic stations,
locations of users, and electricity fees) cannot be estimated, we can use exhaustivity
to list the possibly existent cost models in tests. The model we use here is built on
two assumptions. First, we suppose that the flow loaded on ISPs is proportional to the
overall cost; second, we suppose a unit cost is a function related to the distance. The
study in [26] provides a cost computing scheme, in which the relationship between
equipment operation cost and network scale is logarithmic.

In Chap. 5, we use dynamic game theory to analyze the choice of strategies for
ISPs and users and obtain real user data to verify the model validity.

3.3 Summary

The diversity of the Internet, like the emergence of P2P and mobile Internet, challenges the traditional Internet pricing, which has become researchers' focus of attention.

P2P traffic leads to a cost transfer from CPs to ISPs and users, resulting in unbalanced distribution of profits. So far, researchers have not found a reasonable solution
yet. In Chap. 4, we use cooperative game model to analyze the problem and propose
a profit distribution mechanism based on cooperation between ISPs and CPs, which
is a viable solution this problem.

In mobile Internet, users can choose among various types of services. And the
competition between mobile ISPs is fierce, which provides opportunities for users
to access the Internet via multiple interfaces. In Chap. 5, we use dynamic game
theory to analyze strategies of ISPs to launch service combination packages and
give suggestions for ISPs on the set of the packages based on the results. These

suggestions can help ISPs gain higher profits and promote the healthy development of the mobile Internet market.

References

1. Aggarwal, V., Feldmann, A., Scheideler, C.: Can ISPs and P2P users cooperate for improved performance? ACM SIGCOMM Comput. Commun. Rev. **37**(3), 29–40 (2007)
2. Altman, E., Bernhard, P., Caron, S., et al.: A study of non-neutral networks with usage-based prices. In: Proceedings of the 3rd Workshop on Economic Traffc Management (ETM)'10, Amsterdam, Netherlands, 76–84 (2010)
3. Altman, E., Hanawal, M.K., Sundaresan, R.: Nonneutral network and the role of bargaining power in side payments. In: Proceedings of the 4th Workshop on Network Control and Optimization (NETCOOP)'10, Ghent, Belgium, 66–73 (2010)
4. Altman, E., Legout, A., Xu, Y.: Network non-neutrality debate: an economic analysis. CoRR abs/1012.5862. http://arxiv.org/abs/1012.5862 (2010)
5. Altmanna, J., Chu, K.: How to charge for network services: flat-rate or usage-based? Comput. Netw. **36**(5–6), 519–531 (2001)
6. An-Gang, H. Yong, Guo: From monopoly market to competitive market: the deep social change. Reform Theory **5**, 10–13 (2002)
7. Bakos, Y., Brynjolfsson, E.: Bundling information goods: pricing, profits, and efficiency. Manage. Sci. **45**(12), 1613–1630 (1999)
8. Baumol, W.J., Blinder, A.S., Gale, C.L.: Microeconomics: Principles and Policy. Cincinnati: South-Western Educational Publishing 2001: 212 (2001)
9. Binmore, K., Rubinstein, A., Wolinsky, A.: The nash bargaining solution in economic modelling. RAND J. Econ. **17**, 176–188 (1986)
10. BitTorrent. http://bittorrent.com
11. Cheng, B., Stein, L., Jin, H., et al.: GridCast: improving peer sharing for P2P VoD. ACM Trans. Multimedia Comput. Commun. Appl. **4**(4), 1–31 (2008)
12. Choffnes, D.R., Bustamante, F.: Taming the torrent: a practical approach to reducing cross- ISP traffic in peer-to-peer systems. In: Proceedings of ACM SIGCOMM 2008, Seattle, 363–374 (2008)
13. Chun-Yan, Z., Ke X., Bao-Jin, W., Meng, S.: Optimizing cost and performance in multiple interface mobile hosts. Chin. J. Comput. **34**(11), 2176–2186 (2011)
14. CNNIC. http://www.cnnic.net.cn/hlwfzyj/hlwxzbg/hlwtjbg/201301/t20130115_38508.htm
15. Dimitropoulos, X., Hurley, P., Stoecklin, A.K.M.: on the 95-percentile billing method. In: Proceedings of the 10th Passive and Active Measurement Conference (PAM), Seoul, South Korea, 207–216 (2009)
16. Friedman, J. W.: Oligopoly and the Theory of Games. North-Holland Publisher, Amsterdam and New York (1976)
17. Harrison, A.E.: Productivity, imperfect competition and trade reform: theory and evidence. J. Int. Econ. **36**(1–2), 53–73 (1994)
18. Hei, X., Liang, C., Liu, Y., et al.: A measurement study of a large-scale P2P iptv system. IEEE Trans. Multimedia **9**(8), 1672–1687 (2007)
19. Huan, H. Ke, X. Ying, L.: Internet resource pricing models, mechanisms and methods. Networking Sci. **1**, 48–66 (2012)
20. Huang, Y., Fu, T.Z.J., Chiu, D.M., et al.: Challenges, design and analysis of a large-scale P2P-VoD system. In: Proceedings of ACM SIGCOMM, Seattle, 375–388 (2008)
21. Huang, C., Li, J., Ross, K.W.: Can internet video-on-demand be profitable? In: Proceedings of ACM SIGCOMM 2007, Kyoto, Japan, 133–144 January (2007)
22. Jiang, W., Zhang-Shen, R., Rexford, J., et al.: Cooperative content distribution and traffic engineering in an ISP network. In: Proceedings of ACM SIGMETRICS 2009, Seattle, 239–250 (2009)

23. Karagiannis, T., Rodriguez, P., Papagiannaki, E.: Should internet service providers fear peer-assisted content distribution? In: Proceedings of the 2005 Internet Measurement Conference (IMC), Berkeley (2005)

24. Kaya, A., Chiang, M., Trappe, W.: P2P-ISP cooperation: risks and mitigation in multiple-ISP networks. In: Proceedings of IEEE GLOBECOM 2009, Honolulu, HI, 1–8 (2009)

25. Keon, N., Anandalingam, G.: A new pricing model for competitive telecommunications services using congestion discounts. INFORMS J. Comput. 17(4), 248–262 (2005)

26. Lin, S., Xu, K., Wu, J.P., Wang, N., Zhang, Z., Zhong, Y.F.: Will the three-network convergence happen?—an evolution model based analysis. In: Proceedings of International Conference on Communications in China 2012, 149–154. IEEE (2012)

27. Liu, J. Wang, H.: Understanding peer distribution in the global internet. IEEE Netw. 24(4), 40–44 (2010)

28. Ma, z., Xu, K., Liu, J., Wang, H.: Measurement, modeling and enhancement of bitTorrent-based VoD system. Comput. Netw. 56(3), 1103–1117 (2012)

29. Ma, R.T.B., Chiu, D., Lui, J.C.S., et al.: Internet economics: the use of Shapley value for ISP settlement. In: Proceedings of ACM CoNEXT 2007, New York, 775–787 (2007)

30. Ma, R.T., Chiu, D.M., Lui, J.C.S., Misra, V., Rubenstein, D.: Interconnecting eyeballs to content: a shapley value perspective on ISP peering and settlement. In: Proceedings of the 3rd International Workshop on Economics of Networked Systems, Seattle, 61–66 (2008)

31. Ma, R.T.B., Misra, V., Chiu, D., et al.: On cooperative settlement between content, transit and eyeball internet service providers. In: Proceedings of ACM CoNEXT 2008, 802–815 (2008)

32. Ma, R.T.B., Lee, S.C.M., Liu, J.C.S., et al.: Incentive and service differentiation in P2P networks: a game theoretic approach. IEEE/ACM Trans. Netw. 14(5), 978–991 (2006)

33. Ma, R.T., Chiu, D.M., Lui, J.C.S., Misra, V., Rubenstein, D.: On cooperative settlement between content, transit, and eyeball internet service providers. IEEE/ACM Trans. Netw. 19(3), 802–815 (2011)

34. Maximizing BitTorrent Speeds with uTorrent (Guide/Tutorial) http://www.bootstrike.com/Articles/BitTorrentGuide/

35. Misra, V., Ioannidis, S., Chaintreau, A., et al.: Incentivizing peer-assisted services: a fluid shapley value approach. In: Proceedings of the 2010 ACM SIGMETRICS 2010, 215–226 (2010)

36. Nash, J.F.: The bargaining problem. Econometrica: Int. Econometric Soc. 155–162 (1950)

37. Notz, W.: International private agreements in the form of cartels, cyndicates, and other combinations. J. Polit. Econ. 28(8), 658–679 (1920)

38. Novshek, W., Sonnenschein, H.: General equilibrium with free entry: a synthetic approach to the theory of perfect competition. J. Econ. Lit. 25(3), 1281–1306 (1987)

39. Osborne, M., Rubinstein, A.: A Course in Game Theory. MIT Press, Cambridge (1994)

40. PPLive. http://www.pplive.com

41. PPStream. http://www.ppstream.com

42. Rodriguez, P., Tan, S.M., Gkantsidis, C.: On the feasibility of commercial, legal P2P content distribution. ACM SIGCOMM Comput. Commun. Rev. 36(1), 75–78 (2006)

43. Sen, S., Joe-Wong, C., Ha, S., Chiang, M.: Incentivizing time-shifting of data: a survey of time-dependent pricing for internet access. IEEE Commun. Mag. 50(11), 91–99 (2012)

44. Shen, H.-X., and Basar, T.: Optimal nonlinear pricing for a monopolistic network service provider with complete and incomplete information. IEEE J. Sel. Areas Commun. 25(6), 1216–1223. August 2007

45. Shen, G., Wang, Y., Xiong, Y., et al.: HPTP: relieving the tension between ISPs and P2P. In: Proceedings of the 6th International Workshop on Peer-to-Peer Systems (IPTPS), Bellevue (2007)

46. Suh, K., Diot, C., Kurose, J., et al.: Push-to-peer video-on-demand system: design and evaluation. IEEE J. Sel. Areas Commun. 25(9), 1706–1716 (2007)

47. UUSee. http://www.uusee.com

48. Valancius, V., Laoutaris, N., Massoulie, L., et al.: Greening the internet with nano data centers. In: Proceedings of the ACM CoNEXT 2009, Rome, Italy, 37–48 December 2009

49. Vytautas,V., Cristian, L., Nick, F.: How many tiers? Pricing in the internet transit market. In: Proceedings of the ACM SIGCOMM 2011 Conference, Toronto, ON, Canada, 194–205 (2011)
50. Wang, Q., Chiu, D.M., and Lui, J.C.S.: ISP uplink pricing in a competitive market. In: Proceedings of the International Conference on Telecommunications, St. Petersburg, Russia, 1–6 (2008)
51. Wang, Q., Chiu, D., Lui, J.C.: ISP uplink pricing in a competitive market. In: Proceedings of the 15th Annual International Conference on Telecommunications (ICT)'08, St. Petersburg, Russia, 1–6 (2008)
52. Wang, J.H., Chiu, D.M., Lui, J.C.S.: A game-theoretic analysis of the implications of overlay network traffc on ISP peering. Comput. Netw. **52**(15), 2961–2974 (2008)
53. Wei-Ying, Z.: Game Theory and Information Economics. Shanghai: Shanghai People's Publishing House (2nd edition) (2005)
54. Winter, E.: The Shapley Value. North-Holland: In: Aumzann, R.J., Hart, S., (eds.) The Handbook of Game Theory. pp. 295–303 (2002)
55. Xie, H., Yang, Y.R., Krishnamurthy, A., et al.: P4P: provider portal for applications. In: Proceedings of ACM SIGCOMM 2008, Seattle, 351–362 (2008)
56. Xi-Ren, C., Hong-Xia, S.: Internet pricing with a game theoretical approach: concepts and examples. IEEE/ACM Trans. Netw. **10**(2), 208–216 (2002)
57. Xu, K., Li, H., Liu, J., Zhu, W., Wang, W.: PPVA: A universal and transparent peer-to-peer accelerator for interactive online video sharing. In: Proceedings of IWQoS 2010. 1–9 (2010)
58. Xu, K., Shen, M., Ye, M.: A model approach to estimate peer-to-peer traffic matrices. In: Proceedings of IEEE INFOCOM 2011, 676–684 (2011)
59. Xu, K., Zhong, Y.F., He, H.: Can P2P technology benefit ISPs? A cooperative profit-distribution answer. http://arxiv.org/abs/1212.4915
60. Yaiche, H., Mazumdar, R.R., Rosenberg, C.: Game theoretic framework for bandwidth allocation and pricing in broadband networks. IEEE/ACM Trans. Netw. **8**(5), 667–678 (2000)
61. Yuksel, M., Kalyanarama, S.: Pricing granularity for congestion sensitive pricing. In: Proceedings of Computers and Communications (ISCC 2003), Antalya, Turkey. 169–174 (2003)

Chapter 4
Cooperative Game-Based Pricing and Profit Distribution in P2P Markets

Peer-to-Peer (P2P) technology has been the foundation of many important Internet applications, like Video on Demand (VoD) and file sharing. However, under the traditional pricing mechanism, the fact that most P2P traffic flows among peers can dramatically decrease the profit of ISPs, who may take actions against P2P and impede the development of P2P technology [15, 18]. In this chapter, we develop a mathematical framework to analyze such economic issues. Inspired by the idea from cooperative game theory, we propose a cooperative profit-distribution model based on Nash Bargaining Solution (NBS), in which both eyeball ISPs and Peer-assisted Content Providers (PCPs) form a separate coalition and compute a fair Pareto point to determine profit distribution. Here the eyeball ISPs refer to the ISPs which specialize in delivery to hundreds of thousands of residential users, supporting the last-mile connectivity [8]. Moreover, we design a fair and feasible mechanism for profit distribution within each coalition and give a model to discuss the potential competition among ISPs. We show that such a cooperative method not only guarantees the fair profit distribution among network participators, but also helps improve the economic efficiency of the network system.

4.1 Non-cooperative Game Model

4.1.1 Network Model

The network model consists of three communities: ISP community, CP community, and the user community, which are denoted by M_{ISP}, M_{CP}, and M_{user}, respectively. Their relationships are illustrated in Fig. 4.1. In a practical network system, M_{ISP} often adopts a bandwidth-based pricing model (such as the 95-percentile billing for burstable bandwidth [5]) to charge M_{CP} and a flat pricing model to charge M_{user} [2, 4]. Moreover, M_{CP} often charges M_{user} based on its consumed traffic volume.

K. Xu et al., *Internet Resource Pricing Models*,
SpringerBriefs in Computer Science, DOI: 10.1007/978-1-4614-8409-7_4,
© The Author(s) 2014

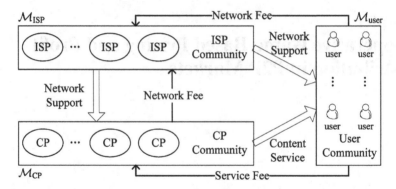

Fig. 4.1 Relationships among ISP community, CP community, and the user community

In the Client/Server (C/S) network, all service contents flow from M_{CP} to M_{user} through M_{ISP}'s network. Suppose the total traversing bandwidth provided by M_{ISP} is b_{ISP}, the bandwidth bought by M_{CP} from M_{ISP} is b_{CP}, and that bought by M_{user} is b_{user}. For M_{CP} and M_{user}, suppose their average bandwidth utilization rates are ξ_{CP} and ξ_{user}, respectively. Usually, ξ_{CP} is high, while ξ_{user} is low (large CPs can use bandwidth more efficiently). Let v be the traffic volume. Then, we have

$$v = b_{CP} \cdot \xi_{CP} = b_{user} \cdot \xi_{user} \tag{4.1}$$

In the peer-assisted network, some CPs in M_{CP} begin to adopt P2P technology and ultimately become Peer-assisted Content Providers (PCPs). After these CPs become PCPs, the CP community becomes $M_{CP} = M_{PCP} \cup M_{CP}^r$, where M_{PCP} is the set of PCPs, and M_{CP}^r consists of the rest unchanged CPs. Then, the service content consists of two parts: the P2P content and the content provided by M_{CP}^r. The former is more complex because it flows to M_{user} from both M_{PCP} and M_{user} through networks provided by ISPs in M_{ISP}. We suppose the traffic of M_{PCP} accounts for a proportion α in the total traffic of M_{CP}. Generally, the P2P content provided by servers of PCPs in M_{PCP} accounts for only a small proportion $\beta > 0$ and the other will be provided by users in M_{user}. In this case, M_{PCP} can reduce its bought bandwidth to a smaller value b_{PCP}^*, so as to reduce the cost and keep its bandwidth utilization rate at ξ_{CP}, while M_{user} with fixed bandwidth at b_{user}, will increase its bandwidth utilization rate to a higher value ξ_{user}^*, which makes the link or path busier. We assume the emergence of P2P traffic will not impact the traffic of M_{CP}^r. Then M_{CP}^r will keep its traffic at $v_{cs} = b_{CP} \cdot (1 - \alpha) \cdot \xi_{CP}$. Suppose the total user-side uploading and downloading P2P traffic amount is $v_{p2p} = \bar{v}_{p2p} + \underline{v}_{p2p}$, where $\bar{v}_{p2p}$ and $\underline{v}_{p2p}$ refer to the user-side uploading and downloading P2P traffic amounts respectively. Then we have the following equations:

$$\underline{v}_{p2p} \cdot \beta = b^*_{PCP} \cdot \xi_{CP}$$
$$\underline{v}_{p2p} \cdot (1 - \beta) = \bar{v}_{p2p}$$

which means that the users' P2P downloading demand with a β proportion is satisfied by the servers of the PCP, and the demand with the other $1 - \beta$ proportion is satisfied by the users themselves. Then we have:

$$\underline{v}_{p2p} = \frac{b^*_{PCP} \cdot \xi_{CP}}{\beta} = \frac{v_{p2p}}{1 + (1 - \beta)}$$

So, $v_{p2p} = b^*_{PCP} \cdot \xi_{CP} \cdot (2 - \beta)/\beta$. Then, similar to the case of C/S network, we should have

$$v_{p2p} + v_{cs} = b^*_{PCP} \cdot \xi_{CP} \cdot \frac{2 - \beta}{\beta} + b_{CP} \cdot (1 - \alpha) \cdot \xi_{CP} = b_{user} \cdot \xi^*_{user} \qquad (4.2)$$

where $\xi_{CP} \geq \xi^*_{user} \geq \xi_{user}$. If $\beta = 1$, CPs will provide the total traffic, which is the same as Eq. (4.1). Here we assume $\beta > 0$, which means the server always provides content and makes the equation meaningful. We also make some assumptions here:

- The ISP charges CP/PCP a unit usage price p_b and charges users a flat fee τ;
- The CP charges users a unit usage price p_s;
- M_{ISP} has built a network with a fixed capacity and it is not fully filled with traffic.

The interaction among M_{ISP}, M_{CP} and M_{user} can be demonstrated by two games, which will be analysed in detail in Sects. 4.1.2 and 4.1.3.

4.1.2 Strategy-Chosen Game

We use a dynamic game between M_{ISP} and M_{CP} to analyze their strategies on technology and pricing. As discussed in Sect. 4.1.1, the CPs can choose between C/S network and P2P-assisted network, and the ISPs have the strategy space of charging users based on flat pricing or usage-based pricing. A dynamic game is used here because the M_{CP} will first choose whether to adopt P2P in content delivery, after which the M_{ISP} will choose the pricing strategy on users according to the strategy of M_{CP}. The game tree of this strategy-chosen game is shown in Fig. 4.2.

As a matter of convenience, we use states 0, 1 and 2 to refer to the possible market states (i.e. the three leaves in the game tree in Fig. 4.2) determined by the strategies chosen by M_{ISP} and M_{CP}, and U^{Si}_{CP} and U^{Si}_{ISP} refer to the profit of M_{CP} and M_{ISP} in State i ($i = 1, 2, 3$), therefore, (U^{Si}_{CP}, U^{Si}_{ISP}) refers to the payoff of the game in each state.

The payoff of M_{ISP} and M_{CP} in each state is determined by the equilibrium of the two-stage price-decision game in Sect. 4.1.3 and the values of the payoffs will determine the equilibrium of this strategy-chosen game.

Fig. 4.2 The strategy-chosen game tree

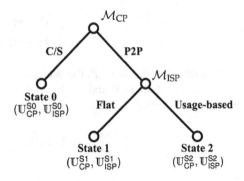

4.1.3 Two-Stage Price-Decision Game

A three-player non-cooperative game can be used to characterize the interactions among M_{ISP}, M_{CP}, and M_{user}. We introduce M_{user} into this game because the users' reactions are involved in the price decision process of M_{ISP} and M_{CP}. The precondition for this game is that both M_{ISP} and M_{CP} have chosen their strategy, which has been discussed in Sect. 4.1.2. We analyze a two-stage game model to determine b_{user}, the bandwidth requirement of M_{user}. We also aim to deduce the basic traffic usage v at equilibrium. The optimal charging fees include ISPs' network fee and CPs' service fee. We use *backward induction* to solve this game and obtain an initial equilibrium market state (State 0).

4.1.3.1 Game Formulation

The strategy spaces of M_{ISP} and M_{CP} are both continuous, so we give an overview rather than the game tree of the two-stage game in Fig. 4.3 because an overview can better demonstrate the strategies of the participators and the repeated game between M_{ISP} and M_{CP}. At the first stage, M_{ISP} and M_{CP} decide the prices through a non-cooperative repeated game; then, at the second stage, M_{user} makes the best response traffic usage decision according to the prices set by M_{ISP} and M_{CP} at the first stage.

Initially, suppose M_{ISP} charges M_{CP} a bandwidth-based price p_b and charges M_{user} a flat price τ. We assume the equivalent bandwidth-based unit price of τ is the same as p_b. Thus, τ is often set based on a given ξ_{user} ($\tau = \frac{v}{\xi_{user}} \cdot p_b$). Then, the profit of M_{ISP} is

$$
\begin{aligned}
\mathbb{U}_{ISP}^{S0}(p_b) &= b_{CP} \cdot p_b + \tau - \mathbb{C}_{ISP}(v) \\
&= \left(\frac{v}{\xi_{CP}} + \frac{v}{\xi_{user}} \right) \cdot p_b - \mathbb{C}_{ISP}(v),
\end{aligned}
\tag{4.3}
$$

where $\mathbb{C}_{ISP}(\cdot)$ is a composite cost function [9].

For M_{CP}, let p_s be unit service price and $\mathbf{F}_{ad}(\cdot)$ be a volume-based advertisement fee function. Then, its profit is

Fig. 4.3 Overview of the two-stage game. The *arrows* illustrate the input and the output of each community and the * represents the final optimal reaction, i.e. the Nash equilibrium

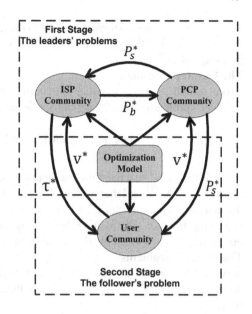

$$U_{CP}^{S0}(p_s) = v \cdot p_s + \mathbf{F}_{ad}(v) - b_{CP} \cdot p_b - \mathbf{C}_{CP}(v)$$

$$= v \cdot p_s + \mathbf{F}_{ad}(v) - \frac{v}{\xi_{CP}} \cdot p_b - \mathbf{C}_{CP}(v), \qquad (4.4)$$

where $\mathbf{C}_{CP}(\cdot)$ is a volume-based cost function.

$\mathbf{U}_{ISP}$ is super additive because the first item on the right side of Eq. (4.3) is linear to the traffic volume and the second item is increasing and concave in the traffic volume, which will be discussed in Sect. 4.1.5, and so is $\mathbf{U}_{CP}$. The super additivity of $\mathbf{U}_{ISP}$ and $\mathbf{U}_{CP}$ guarantees that the ISP grand coalition and the PCP grand coalition can be formed, because with more participators joining in, the total profit will increase accordingly [10].

In addition, let $\mathbf{E}_{user}(v)$ be the experience value for $\mathbf{M}_{user}$ consuming content volume v. Then, its utility is

$$U_{user}^{S0}(v) = \mathbf{E}_{user}(v) - (b_{user} \cdot p_b + v \cdot p_s)$$

$$= \mathbf{E}_{user}(v) - \left(\frac{p_b}{\xi_{user}} + p_s \right) \cdot v. \qquad (4.5)$$

In this C/S network, a three-player game can characterize the interactions. $\mathbf{M}_{ISP}$ and $\mathbf{M}_{CP}$ act as leaders to price $\mathbf{M}_{user}$ which acts as a follower to decide traffic usage. In addition, since $\mathbf{M}_{ISP}$ and $\mathbf{M}_{CP}$ jointly affect the resource usage of $\mathbf{M}_{user}$, a two-player non-cooperative game happens between them.

According to the *backward induction* in the leader-follower game, we first analyze the second stage of this game, assuming that M_{ISP} and M_{CP} have set the prices at the first stage of the game.

The follower's problem

Given p_b and p_s, M_{user} is going to maximize its utility in Eq. (4.5). By solving the follower's problem (Stage 2), we can obtain the service content consumed by M_{user} as

$$\hat{v}(p_b, p_s) = \min\{\underset{v}{\arg\max}\, \mathbb{U}_{user}, b_{user} \cdot \xi_{cp}\}. \tag{4.6}$$

which is the users' best response traffic usage decision within its purchased capacity. Let $o(v) = \frac{d\mathbf{E}_{user}(v)}{dv}$ and then we have $o(v) = d \cdot p_b + p_s$ ($d = 1/\xi_{user}$) according to the first order condition for Eq. (4.5). As $d \cdot p_b + p_s > 0$, $\mathbf{E}_{user}(v)$ is a continuously increasing function. We assume that $o(v)$ is a one-to-one mapping. Then based on our assumption that the network is underused, we have $\hat{v}(p_b, p_s) = o^{-1}(d \cdot p_b + p_s)$.

The leaders' problems

Suppose M_{ISP} and M_{CP} know their impacts on the utility of M_{user}. Then, anticipating users will choose $v = \hat{v}(p_b, p_s)$ to decide their traffic usage; the leaders' problems become

$$\text{For } M_{ISP}: \quad \underset{pb}{\max}\, \mathbb{U}_{ISP}(p_b, \hat{v}(p_b, p_s))$$

$$\text{For } M_{CP}: \quad \underset{ps}{\max}\, \mathbb{U}_{CP}(p_s, \hat{v}(p_b, p_s)).$$

Then a two-player non-cooperative Nash game between M_{ISP} and M_{CP} happens. M_{ISP} and M_{CP} take turns to optimize their own objects $\mathbb{U}_{ISP}$ and $\mathbb{U}_{CP}$ by varying their own decision variables p_b and p_s, respectively, while keeping that of the other player as a constant. The existence of Nash Equilibrium (NE) for this multi-leader-follower game depends on the properties of each utility function and the existence and the uniqueness of pure Nash equilibrium have been well proved for particular continuous Nash game [3].

4.1.3.2 Game Solution

Let (p_b^*, p_s^*) be the Nash Equilibrium. Then, according to the definition of Nash Equilibrium, the solution turns out to be as follows

$$\begin{cases} p_b^* = \underset{p_b}{\arg\max}\, \mathbb{U}_{ISP}(p_b, \hat{v}(p_b, p_s^*)), \\ p_s^* = \underset{p_s}{\arg\max}\, \mathbb{U}_{CP}(p_b^*, \hat{v}(p_b^*, p_s)). \end{cases} \tag{4.7}$$

We have the following theorem on the simplified sufficient conditions of Nash Equilibrium for this problem, which will also help us obtain equilibrium.

Theorem 4.1 *Let* (p_b^*, p_s^*) *be the Nash Equilibrium defined in Eq.(4.7) and* $v^* = \hat{v}(p_b^*, p_s^*)$. *Then, let*

$$\phi_1(v) = c \cdot v \cdot \frac{1}{d} \cdot \frac{do(v)}{dv} - \frac{dC_{ISP}(v)}{dv},$$

$$\phi_2(v) = v \cdot \frac{do(v)}{dv} + \frac{dF_{ad}(v)}{dv} - \frac{dC_{CP}(v)}{dv}.$$

$$(4.8)$$

It must satisfy the following two conditions:

1. $o(v^*) + \phi_1(v^*) + \phi_2(v^*) = 0$,
2. $(\frac{c}{d} \cdot \frac{do(v)}{dv} + \frac{d\phi_1(v)}{dv})|_{v^*} < 0$, $(\frac{do(v)}{dv} + \frac{d\phi_2(v)}{dv})|_{v^*} < 0$.

where $c = \frac{1}{\xi_{user}} + \frac{1}{\xi_{cp}}$, *and* $e = \frac{1}{\xi_{cp}}$.

Proof. We have $o(v^*) = d \cdot p_b^* + p_s^*$ according to Eq.(4.5).

According to the definition of Nash Equilibrium, p_b^* should be the best response to p_s^*, and vice versa. Since we do not consider the cases where the maximum profit happens at the boundary, we must have

$$\frac{\partial E_{ISP}(p_b, p_s^*)}{\partial p_b}|_{p_b^*} = 0, \quad \frac{\partial E_{ISP}^2(p_b, p_s^*)}{\partial p_b}|_{p_b^*} < 0,$$

$$\frac{\partial E_{CP}(p_b^*, p_s)}{\partial p_s}|_{p_s^*} = 0, \quad \frac{\partial E_{CP}^2(p_b^*, p_s)}{\partial p_s}|_{p_s^*} < 0.$$

$$(4.9)$$

Moreover, because $v = o^{-1}(d \cdot p_b + p_s)$ exists, we have

$$\frac{\partial E_{ISP}(v, p_s^*)}{\partial v}|_{v^*} = \frac{\partial E_{ISP}(p_b, p_s^*)}{\partial p_b}|_{p_b^*} \cdot \frac{\partial p_b}{\partial v}|_{v^*}, \quad \frac{\partial p_b}{\partial v}|_{v^*} = \frac{1}{d}\frac{do(v)}{dv}|_{v^*} \neq 0,$$

$$\frac{\partial E_{CP}(v, p_b^*)}{\partial v}|_{v^*} = \frac{\partial E_{ISP}(p_b^*, p_s)}{\partial p_s}|_{p_s^*} \cdot \frac{\partial p_s}{\partial v}|_{v^*}, \quad \frac{\partial p_s}{\partial v}|_{v^*} = \frac{do(v)}{dv}|_{v^*} \neq 0.$$

We can apply the above properties under conditions in Eq.(4.9) and rewrite them as follows:

$$\frac{\partial E_{ISP}(v, p_s^*)}{\partial v}|_{v^*} = 0, \quad \frac{\partial E_{ISP}^2(v, p_s^*)}{\partial v}|_{v^*} < 0,$$

$$\frac{\partial E_{CP}(p_b^*, v)}{\partial v}|_{v^*} = 0, \quad \frac{\partial E_{CP}^2(p_b^*, v)}{\partial v}|_{v^*} < 0.$$

$$(4.10)$$

From Eqs.(4.10), (4.3), (4.4) and $o(v^*) = d \cdot p_b^* + p_s^*$, we derive that

$$c \cdot p_b^* + c \cdot v^* \cdot \frac{1}{d} \cdot \frac{do(v)}{dv}|_{v^*} - \frac{dC_{ISP}(v)}{dv}|_{v^*} = 0,$$

$$p_s^* + v^* \cdot \frac{do(v)}{dv}|_{v^*} + \frac{dF_{ad}(v)}{dv}|_{v^*} - e \cdot p_b - \frac{dC_{CP}(v)}{dv}|_{v^*} = 0.$$

It can be further simplified based on Eq.(4.8) as follows

$$c \cdot p_b^* + \phi_1(v^*) = 0,$$

$$p_s^* + \phi_2(v^*) - e \cdot p_b^* = 0.$$

$$(4.11)$$

v^* must satisfy the following condition in order to be the traffic usage at a Nash Equilibrium

$$o(v^*) + \phi_1(v^*) + \phi_2(v^*) = 0. \tag{4.12}$$

Second, order conditions in Eq. (4.10) can be simplified as

$$(\frac{c}{d} \cdot \frac{\partial o(v)}{\partial (v)} + \frac{\partial \phi_1(v^*)}{\partial v})|_{v^*} < 0, \quad (\frac{\partial o(v)}{\partial (v)} + \frac{\partial \phi_2(v^*)}{\partial v})|_{v^*} < 0. \tag{4.13}$$

This theorem presents a way to compute the Nash Equilibrium of the game which represents the steady state of this network market (denoted as State 0).

4.1.4 P2P-Involved Profit Computing Model

One important job of this chapter is to measure and quantify P2P traffic's impact on the network economic market under the traditional pricing mechanism. Based on results in the last subsection, we first analyze the growing impact of P2P traffic on the profits or utilities of Internet participators when the pricing strategy remains unchanged, which we define as State 1. Then we illustrate an analysis of M_{ISP}'s reactive behavior conditionally and study its corresponding aftermath, i.e., State 2. Finally, we present a state transformation graph to summarize these possible non-cooperative market states and their transition conditions.

4.1.4.1 State 1

In the peer-assisted network, we assume v_{cs} will not be impacted by the emergence of P2P traffic here (i.e. $v_{cs} = v_{cs}^* = v^* \cdot (1 - \alpha)$). It is reasonable when people give priority to satisfying inelastic basic needs of traditional Internet services (such as email and web) which are unlikely to become P2P-assisted.

Compared with C/S content distribution mode, P2P can improve the experience of M_{user} because of its scalability. So let $\widehat{E}_{user}$ be M_{user}'s new experience value for content downloading profile $v = v_{p2p} + v_{cs}^*$, and we assume $\widehat{E}_{user}(v) > E_{user}(v)$ as long as $v > v_{cs}^*$ (i.e., $v_{p2p} > 0$). Let a be the experience accelerating factor of P2P traffic (which is related to β and always satisfies $a > 1$), so we have $\widehat{E}_{user}(v) = E_{user}(a \cdot (v - v_{cs}^*) + v_{cs}^*)$. Indeed, we simply assume that a and β satisfy a linear relationship. So we create fitting curve for a based on two empirical points $(\beta, a) = (0.3, 4)$ (i.e., when 70% PCP content is provided by P2P, users' experience will expand four times compared with that under C/S mode) and $(\beta, a) = (1, 1)$ (i.e., when all the PCP content is provided by servers, the calculation of such experience is the same as that under C/S mode). Then, we get $a = 1 + \frac{30}{7}(1 - \beta)$.

Remark 4.1 Intuitively, $1 - \beta$ reflects P2P's power, and when it becomes larger, the performance of P2P service will become better because of its distributed sharing nature. So we assume k increases in accordance with $1 - \beta$. As PCPs' servers guarantee system stability, they are generally indispensable (i.e., $\beta > 0$).

Often, M_{ISP} charges M_{user} a flat price. Suppose new average bandwidth utilization rate ξ_{user}^* cannot exceed ξ_{CP} as discussed in Sect. 4.1.1. Then, we have $v_{p2p} \cdot (2 - \beta) + v_{cs}^* \le b_{user}^* \cdot \xi_{CP}$. Let $\widetilde{v}_{p2p} = \frac{b_{user}^* \cdot \xi_{CP} - v_{cs}^*}{2 - \beta}$. As long as $v_{p2p} \le \widetilde{v}_{p2p}$, the fee charged from M_{user} will be kept at $\tau = b_{user}^* \cdot p_b^*$; when $v_{p2p} > \widetilde{v}_{p2p}$, we assume M_{ISP} will charge additional fee for the excessive volume $(v_{p2p} - \widetilde{v}_{p2p}) \cdot (2 - \beta)$ based on a volume-based pricing. For bandwidth-based price p_b^*, its equivalent volume-based price is $\frac{p_b^*}{\xi_{user}}$. Thus, the utility of M_{user} becomes

$$
\mathbb{U}_{user}^{S1} = \begin{cases} \widehat{\mathbb{E}}_{user}(v) - v \cdot p_s^* - \tau, & \text{if } v_{p2p} \le \widetilde{v}_{p2p}; \\ \widehat{\mathbb{E}}_{user}(v) - v \cdot p_s^* - \tau - (v_{p2p} - \widetilde{v}_{p2p}) \cdot (2 - \beta) \cdot \frac{p_b^*}{\xi_{user}}, & \text{otherwise.} \end{cases}
$$
(4.14)

Here, M_{user} will decide v_{p2p}^{S1} (since $v = v_{p2p} + v_{cs}^*$ based on our assumption) to maximize $\mathbb{U}_{user}$, i.e.,

$$
v_{p2p}^{S1} = \operatorname*{argmax}_{v_{p2p}} \mathbb{U}_{user}.
$$
(4.15)

Then, based on v_{p2p}^{S1}, we can get $\mathbb{U}_{CP}$ and $\mathbb{U}_{ISP}$ as follows.

For M_{CP}, $\mathbb{U}_{CP}$ will become

$$
\mathbb{U}_{CP}^{S1} = v^{S1} \cdot p_s^* + \mathbf{F}_{ad}(v^{S1}) - \frac{v_{p2p}^{S1} \cdot \beta + v_{cs}^*}{\xi_{CP}} \cdot p_b^* - \widehat{\mathbf{C}}(v^{S1}),
$$
(4.16)

where $v^{S1} = v_{p2p}^{S1} + v_{cs}^*$, and $\frac{v_{p2p}^{S1} \cdot \beta + v_{cs}^*}{\xi_{CP}}$ denotes the bandwidth purchased by M_{CP} from M_{ISP} when the β proportion traffic is provided by their own servers. Similar to $\widehat{\mathbf{E}}_{user}(v)$, here we define $\widehat{\mathbf{C}}_{CP}(v) = \mathbf{C}_{CP}((v - v_{cs}^*) \cdot \beta + v_{cs}^*)$ $(0 < \beta \le 1)$ to measure the cost alleviated by P2P-assisting.

Accordingly, $\mathbb{U}_{ISP}$ will become

$$
\mathbb{U}_{ISP}^{S1} = \begin{cases} \tau + \frac{v_{p2p}^{S1} \cdot \beta + v_{cs}^*}{\xi_{CP}} \cdot p_b^* - \mathbf{C}_{ISP}(v^{S1}), & \text{if } v_{p2p}^{S1} \le \widetilde{v}_{p2p}; \\ \tau + (v_{p2p}^{S1} - \widetilde{v}_{p2p}) \cdot (2 - \beta) \cdot \frac{p_b^*}{\xi_{user}} + \frac{v_{p2p}^{S1} \cdot \beta + v_{cs}^*}{\xi_{CP}} \cdot p_b^* - \mathbf{C}_{ISP}(v^{S1}), & \text{otherwise.} \end{cases}
$$
(4.17)

4.1.4.2 State 2

For M_{ISP}, one main reason for the decrease of its profit is that it charges M_{user} a flat price, which leads to P2P free-riding. To defeat such free-riders, one effective way is to change the original flat pricing model into a volume-based pricing model [6, 14, 16]. Like in State 1, we adopt $\frac{p_b^*}{\xi_{user}}$ as the volume-based price. Then, the utility of M_{user} becomes

$$\mathbb{U}_{\text{user}}^{\text{S2}} = \widehat{\mathbf{E}}_{\text{user}}(v) - v \cdot p_s^* - \left[v_{\text{p2p}} \cdot (2 - \beta) + v_{\text{cs}}^*\right] \cdot \frac{p_b^*}{\xi_{\text{user}}}. \qquad (4.18)$$

Similar to Eq. (4.15), M_{user} chooses

$$v_{\text{p2p}}^{\text{S2}} = \min\{\underset{v}{\text{argmax}}\,\mathbb{U}_{\text{user}}, \widetilde{v}_{\text{p2p}}\} \qquad (4.19)$$

to obtain the feasible optimal traffic usage. Then, the utilities of M_{ISP} and M_{CP} can thus be obtained. For $\mathbb{U}_{\text{CP}}$, the computation method is equal to Eq. (4.16). Accordingly, $\mathbb{U}_{\text{ISP}}$ becomes

$$\mathbb{U}_{\text{ISP}}^{\text{S2}} = \left[v_{\text{p2p}}^{\text{S2}} \cdot (2 - \beta) + v_{\text{cs}}^*\right] \cdot \frac{p_b^*}{\xi_{\text{user}}} + \frac{v_{\text{p2p}}^{\text{S2}} \cdot \beta + v_{\text{cs}}^*}{\xi_{\text{CP}}} \cdot p_b^* - \mathbf{C}_{\text{ISP}}(v^{\text{S2}}) \qquad (4.20)$$

where $v^{\text{S2}} = v_{\text{p2p}}^{\text{S2}} + v_{\text{cs}}^*$.

4.1.4.3 Discussion and Non-cooperative State Analysis

In this subsection, we describe another two possible non-cooperative states. They help us analyze *how P2P technology will affect the network participators' behaviors and utilities*. Through analyzing these states, we can quantify the profits and predict the possible profit changing trends of our network participators under fixed traffic profile and at unchanged pricing levels. The reasons are as follows:

(a) We need to study how P2P traffic impacts profit distribution among these players if non-P2P traffic is treated in the same way that it is treated at State 0 by M_{ISP} and M_{user};
(b) While minimizing its cost, M_{CP}^r decides β mostly based on its own technology and network situation, rather than the complex economic computation.

As shown in Fig. 4.4, we summarize the state transformation conditions among States 0, 1, and 2. Unlike the way we analyze dynamic games of complete information by using *game trees* directly [13], we summarize all possible equilibrium states (i.e., subgame perfect Nash equilibriums, SPNEs) which the system can attain and the conditions under which each state is the SPNE. The state transmission here specifies that in practical networks, the proper Nash equilibrium may not be reached according to analysis and prediction, but may be attained through several steps of state transformations. For example, as pricing strategies act as long-term behaviors of M_{ISP}, it cannot be dynamic and flexible because of traffic caused by M_{CP}. Thus,

Fig. 4.4 State transformation among States 0, 1, and 2

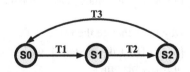

after the system passes through a long path, it is likely to arrive at a reasonable Nash equilibrium finally.

4.1.5 Examples and Analysis

We note that in a practical system, the cost function $C_{CP}(\cdot)$ is often an increasing and concave function on variable v, and $C_{CP}(0) = c_{CP}$ (where c_{CP} is a constant). While $C_{ISP}(\cdot)$ is often a continuously increasing function on variable v with a fixed cost $C_{ISP}(0) = c_{ISP}$. When v is small, the growth rate of this cost decreases with a larger v, while when v is large, the growth rate of this cost increases with an even larger v due to congestion. Generally, the two functions $F_{ad}(\cdot)$ and $E_{user}(\cdot)$ are increasing and concave in accordance with $F_{ad}(0) = 0$ and $E_{user}(0) = 0$.

The changing trends of $C_{CP}(v)$, $C_{ISP}(v)$, and $E_{user}(v)$ is shown in Fig. 4.5, which has the above mentioned properties. While many forms are reasonable, we take $C_{CP}(v) = \ln(v+1) + 0.2$ to model the cost of M_{CP}. We use congestion cost, $C_{ISP}(v) = \ln(v+1) + 100(\frac{1}{b_{ISP}-v} - \frac{1}{b_{ISP}}) + 0.4$, to indicate potential expansion cost for ISPs, which will increase fast when v approaches to b_{ISP} [9]. We also take simple forms for $F_{ad}(v) = 5\ln(v+1)$ and $E_{user}(v) = 5\ln(v+1)$, to satisfy the increasing and concave property. Then, according to Eq. (4.6) and our assumption, we have $\hat{v}(p_b, p_s) = \frac{5}{d \cdot p_b + p_s} - 1$ which will not exceed the capacity. Based on Theorem 4.1, we can directly derive the Nash Equilibrium in closed-form. Then, we can further study the sensitivity of v^* on variables ξ_{CP} and ξ_{user}. Here, we just assume $b_{ISP} = 100$ because the bandwidth can scale under the previously mentioned assumption without affecting the results; we also assume that $0.1 \leq \xi_{user}, \xi_{CP} \leq 0.75$ and $\xi_{user} \leq 0.4 \cdot \xi_{CP}$ based on experience, and we find that the smaller ratio of ξ_{user} to ξ_{CP}, the higher v^* becomes.

Fig. 4.5 Changing trends of $C_{CP}(v)$, $C_{ISP}(v)$, and $E_{user}(v)$

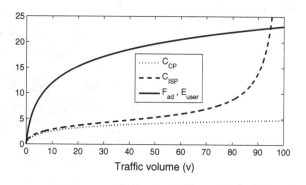

Traffic volume (v)

4.1.5.1 State 0

Similar to Norton's prediction on bandwidth utilization rates [12], we assume $\xi_{CP} = 0.75$, and $\xi_{user} = 0.25$. Then we have $\hat{v}(p_b, p_s) = \frac{5}{4 \cdot b_p + p_s} - 1$, which will not exceed the capacity. Thus the unique Nash Equilibrium point can be computed as where $p_b^* = 0.3321$ ($\tau = 2.8607$), $p_s^* = 0.2571$, $b_{CP} = 2.8713$ and $b_{user} = 8.6140$. The corresponding utilities of each network participator are $(U_{ISP}^{S0}, U_{CP}^{S0}, U_{user}^{S0}) = (2.2438, 3.9942, 2.3281)$.

4.1.5.2 State 1

Following the foregoing example, we take $\alpha = 0.6$, $\beta = 0.3$ as previously assumed. Then by solving the user-side optimization problem in Eq. (4.15), we can deduce the optimal point v_{p2p}^{S1}. It is clear that $v_{p2p}^{S1} > v^* \cdot \alpha$. In addition, we can see that v_{p2p}^{S1} increases to but does not exceed $\tilde{v}_{p2p}$. This implies that M_{user} will try to use up its original bandwidth bought from M_{ISP} with a flat price but without purchasing additional bandwidth. Then, we can obtain $(U_{ISP}^{S1}, U_{CP}^{S1}, U_{user}^{S1}) = (1.5964, 7.2021, 9.6230)$ for State 1. Compared with State 0, U_{CP} increases by 80.31 %, while U_{ISP} decreases by 28.85 %. Thus, motivated by profit increase, some CPs will adopt P2P technology and become PCPs. Then, the overall system will change from State 0 to State 1.

Remark 4.2 Economically, the only condition for the system to change from State 0 to State 1 is that under the traditional pricing mechanism, $U_{CP}^{S1} > U_{CP}^{S0}$. According to Eqs. (4.5) and (4.14), it is easily proved that $v_{p2p}^{S1} + v_{cs}* > v^*$ is always true (See Fig. 4.6).

Fig. 4.6 Traffic volume (v) in States 0, 1, and 2 for different α and β. Note that $v = v_{p2p} + v_{cs}^*$ in States 1 and 2

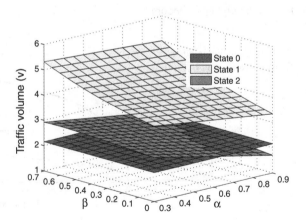

4.1.5.3 State 2

In the previous example, the optimal v_{p2p} sharply decreases to $v_{p2p}^{S2} = 2.3838$. Correspondingly, we have $(\mathbb{U}_{ISP}^{S2}, \mathbb{U}_{CP}^{S2}, \mathbb{U}_{user}^{S2}) = (3.5180, 5.6450, 5.1712)$. Therefore, after M_{ISP} adopts a volume-based pricing model, $\mathbb{U}_{ISP}$ increases by 120.66 %, while $\mathbb{U}_{CP}$ decreases by 29.42 %. Thus, motivated by profit increase, M_{ISP} will change its flat pricing model on M_{user} into a volume-based one. Then, the overall system will change from State 1 to State 2. Since $\mathbb{U}_{CP}^{S2} > \mathbb{U}_{CP}^{S0}$, M_{PCP} still benefits from P2P technology and will not take further actions against M_{ISP} except for a better choice.

Remark 4.3 The two conditions for the overall system to change from State 1 to State 2 are $\mathbb{U}_{ISP}^{S1} < \mathbb{U}_{ISP}^{S0}$ and $\mathbb{U}_{ISP}^{S2} > \mathbb{U}_{ISP}^{S1}$, respectively. For the first one: If $\mathbb{U}_{ISP}^{S1} > \mathbb{U}_{ISP}^{S0}$, M_{ISP} will benefit from P2P technology. However, according to Eqs. (4.14) and (4.18), it is easy to prove that $v_{p2p}^{S2} < v_{p2p}^{S1}$ is always true. Then, M_{ISP} does not need to change its pricing model on M_{user}.

Remark 4.4 For M_{PCP}, if $\mathbb{U}_{CP}^{S2} < \mathbb{U}_{CP}^{S0}$ (Since traffic demand is suppressed in M_{ISP}'s user-side new pricing mode, the saved cost cannot offset the reduced income), it may give up P2P technology because of the reduced profit. Then, the overall system will be forced to change from State 2 to State 0.

4.1.5.4 Analysis

The *game tree* of this example can be illustrated by Fig. 4.2. As the tree shows, the game starts from the M_{CP}'s decision of whether to adopt P2P technology or not. The capacity of P2P is externally decided by the network. If M_{CP} adopts P2P, the game then goes to the M_{ISP}'s decision of which pricing model will be used to charge M_{user}, i.e., flat or usage-based. Once M_{ISP} makes its choice, the game is over. Based on *backward induction* and the payoff results given in this example, we get (P2P, usage-based pricing) as the SPNE, and the equilibrium payoff vector is (5.0835, 3.5226). We can verify that it satisfies the conditions for State 2 to be the final state (i.e., T1 instead of T2 in Fig. 4.4).

In a practical system, since the implementation of network pricing lags behind the technology application, $\mathbb{U}_{CP}^{S1} > \mathbb{U}_{CP}^{S0}$ is always true. Some examples are illustrated in Fig. 4.7b. Thus, the overall system will always change from State 0 to State 1. If $\mathbb{U}_{ISP}^{S1} \geq \mathbb{U}_{ISP}^{S0}$, which only applies to large β in Fig. 4.7a, and M_{ISP} predicts $\mathbb{U}_{ISP}^{S1} \geq \mathbb{U}_{ISP}^{S2}$, the system will stay in State 1. Then, only when $\mathbb{U}_{CP}^{S1} \geq \mathbb{U}_{CP}^{S0}$, the system will stop in State 1 (i.e., the SPNE); Otherwise, if $\mathbb{U}_{ISP}^{S1} < \mathbb{U}_{ISP}^{S0}$ and $\mathbb{U}_{ISP}^{S2} > \mathbb{U}_{ISP}^{S0}$ (as shown in Fig. 4.7a), it will change from State 1 to State 2. Then, if $\mathbb{U}_{CP}^{S2} > \mathbb{U}_{CP}^{S0}$ (as shown in Fig. 4.7b), the system will stop in State 2. Otherwise it will change from State 2 to State 0 and finally stop in State 0. Therefore, by using the state transformation conditions in Fig. 4.4, we can conclude the conditions for each SPNE. Under a certain condition, each accepting state can be a proper Nash equilibrium.

For different traffic profiles (α, β), by solving optimization problems (based on Eqs. (4.15) and (4.19)) of M_{user}, we get the optimal traffic usage with "flat" and "usage-based" pricing strategies of M_{ISP}. Then, according to Eqs. (4.16), (4.17) and (4.20), we can correspondingly derive the utilities of M_{ISP} and M_{CP} as Fig. 4.7 shows. We plot the initial equilibrium utilities computed in Sect. 4.1.3 as a comparison.

As mentioned in Sect. 4.1.1, β is practically small, so we assume that β is smaller than 0.5. Then, through examples illustrated in Fig. 4.7 and based on the above conditions, we can predict that the overall system will finally stop in State 2, where M_{ISP} charges M_{user} with a usage-based pricing model. Here, U_{ISP}^{S2} is 120.66 % more than U_{ISP}^{S1} and 56.99 % more than U_{ISP}^{S0}; U_{CP}^{S2} is 29.42 % less than U_{CP}^{S1} though it is 27.27 % more than U_{CP}^{S0}.

4.2 Cooperative Profit Distribution Model

In this section, we propose a cooperative profit distribution model based on the concept of Nash Bargaining Solution (NBS) [11], in which eyeball ISPs and PCPs first form two coalitions to cooperatively maximize their total profit and then fairly distribute profit based on NBS.

4.2.1 Profit Distribution Between ISP Coalition and PCP Coalition

According to our analysis in Sect. 4.1.4.1, we notice that in the peer-assisted network, M_{user} may use up its original bandwidth bought from M_{ISP} with a flat price without buying additional bandwidth at a volume-based price. Here we consider the following cooperation: PCP coalition sells content at a discount rate γ_{PCP} and ISP coalition charges the increased bandwidth bought by M_{user} at a discount rate γ_{ISP} (where

(a)   **(b)**

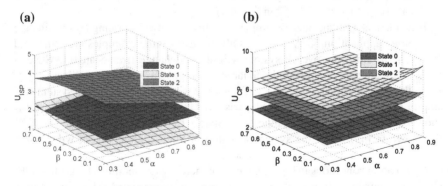

Fig. 4.7 **a** U_{ISP} and **b** U_{CP} for different α and β

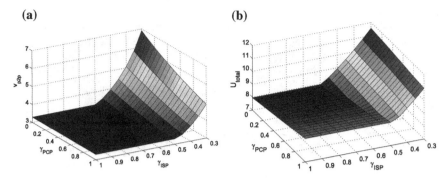

Fig. 4.8 **a** v_{p2p} and **b** U_{total} for different γ_{ISP} and γ_{PCP} in the peer-assisted network with traffic profiles $(\alpha, \beta) = (0.6, 0.3)$

$0 \leq \gamma_{PCP}, \gamma_{ISP} \leq 1$). Both of them try to incentivize M_{user} to consume more content and to buy more bandwidth for P2P applications. It should be noted that if γ_{ISP} is large, v_{p2p} will not increase even if $\gamma_{PCP} = 0$ (See Fig. 4.8). It implies that without the cooperation of ISP coalition, PCP coalition cannot unilaterally incentivize M_{user} to consume more P2P content, and thus the total profit cannot be increased. For PCP coalition, besides the fee charged by ISP coalition according to its direct traffic volume $v \cdot \beta$, some of its profit should be shared with ISP coalition.

In this cooperation, the utility of M_{user} becomes

$$U_{user} = \begin{cases} \widehat{E}_{user}(v) - (v_{p2p} \cdot \gamma_{PCP} + v_{cs}^*) \cdot p_s^* - \tau, & \text{if } v_{p2p} \leq \widetilde{v}_{p2p} \\ \widehat{E}_{user}(v) - (v_{p2p} \cdot \gamma_{PCP} + v_{cs}^*) \cdot p_s^* - \tau - \\ (v_{p2p} - \widetilde{v}_{p2p}) \cdot (2 - \beta) \cdot \frac{p_b^*}{\xi_{user}} \cdot \gamma_{ISP}, & \text{otherwise.} \end{cases}$$

Accordingly, U_{ISP} will become

$$U_{ISP} = \begin{cases} \tau + \frac{v_{p2p} \cdot \beta + v_{cs}^*}{\xi_{CP}} \cdot p_b^* - C_{ISP}(v), & \text{if } v_{p2p} \leq \widetilde{v}_{p2p}; \\ \tau + (v_{p2p} - \widetilde{v}_{p2p}) \cdot (2 - \beta) \cdot \frac{p_b^*}{\xi_{user}} \cdot \gamma_{ISP} + \\ \frac{v_{p2p} \cdot \beta + v_{cs}^*}{\xi_{CP}} \cdot p_b^* - C_{ISP}(v), & \text{otherwise.} \end{cases}$$

Also, U_{CP} will become

$$U_{CP} = (v_{p2p} \cdot \gamma_{PCP} + v_{cs}^*) \cdot p_s^* + F_{ad}(v) - \frac{v_{p2p} \cdot \beta + v_{cs}^*}{\xi_{CP}} \cdot p_b^* - \widehat{C}_{CP}(v).$$

Here, a leader–follower game happens between the cooperative group and M_{user}. The former changes γ_{ISP} and γ_{PCP} to maximize its total profit:

$$U_{total} = U_{ISP} + U_{CP}.$$

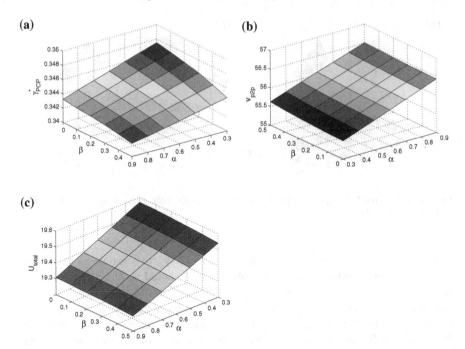

Fig. 4.9 Optimal v_{p2p}, γ_{PCP} and $\mathbb{U}_{total}$ for different α and β

While M_{user} as the price taker changes v_{p2p} to maximize $\mathbb{U}_{user}$:

$$\text{for the } M_{user}:$$
$$\widehat{v} = \underset{v_{p2p}}{\text{argmax}}\, \mathbb{U}_{user}(\gamma_{ISP}, \gamma_{PCP});$$
$$\text{for the cooperative group:} \tag{4.21}$$
$$\max_{\gamma_{ISP}, \gamma_{PCP}} \mathbb{U}_{total}(\gamma_{ISP}, \gamma_{PCP}, \widehat{v}(\gamma_{ISP}, \gamma_{PCP})).$$

By solving the above leader–follower problem under different traffic profiles, we obtain the optimal values of v_{p2p}, γ_{PCP}, and $\mathbb{U}_{total}$ illustrated in Fig. 4.9. From Fig. 4.9, we can observe that γ_{PCP} and $\mathbb{U}_{total}$ decrease accordingly with the increase of α or β, while v_{p2p} increases with the increase of α or β.

Then, for traffic profile $(\alpha, \beta) = (0.6, 0.6)$, we can obtain the unique Stacklberg Equilibrium point where $\gamma_{ISP}^* = 0$, $\gamma_{PCP}^* = 0.3443$, and $v_{p2p}^{S3} = 56.0140$. The results indicate that M_{ISP} will freely upgrade M_{user}'s access bandwidth. Correspondingly, $(\mathbb{U}_{total}^{S3}, \mathbb{U}_{user}^{S3}) = (19.4287, 19.0598)$. We can see that after ISP and PCP coalitions cooperate with each other, both $\mathbb{U}_{total}$ and $\mathbb{U}_{user}$ increase dramatically. Before PCP coalition shares some profit with ISP coalition, $(\mathbb{U}_{ISP}^{S3'}, \mathbb{U}_{CP}^{S3'}) = (4.90593, 14.5227)$. For all cases, $\mathbb{U}_{ISP} + \mathbb{U}_{CP} \leq \mathbb{U}_{total}^{S3}$. Thus,

$$\mathbb{U}_{ISP} + \mathbb{U}_{CP} = \mathbb{U}_{total}^{S3} \qquad (4.22)$$

is the corresponding Pareto boundary.

Now, we are facing an important question: *How can ISP and PCP coalitions choose a fair point on the Pareto boundary as their profit distribution?* As discussed previously, without cooperation, their profit distribution may reach one of the following points (see Fig. 4.4): $(\mathbb{U}_{ISP}^{S0}, \mathbb{U}_{CP}^{S0})$, $(\mathbb{U}_{ISP}^{S1}, \mathbb{U}_{CP}^{S1})$, or $(\mathbb{U}_{ISP}^{S2}, \mathbb{U}_{CP}^{S2})$. In Nash bargaining, such a point is called the *starting point* [1]. If no agreement can be reached by the two bargainers, the starting point will be the outcome of the game. We denote it as $(\mathbb{U}_{ISP}^{s}, \mathbb{U}_{CP}^{s})$. Then, according to the fairness concept of NBS, the fair profit distribution will be on the Pareto boundary and can be deduced by

$$\begin{aligned} \underset{\mathbb{U}_{ISP}, \mathbb{U}_{CP}}{\text{maximize}} \ & (\mathbb{U}_{ISP} - \mathbb{U}_{ISP}^{s})(\mathbb{U}_{CP} - \mathbb{U}_{CP}^{s}), \\ \text{subject to} \ & \mathbb{U}_{ISP} + \mathbb{U}_{CP} = \mathbb{U}_{total}^{S3}. \end{aligned} \qquad (4.23)$$

Here, NBS satisfies all the following four axioms [1, 11, 19]: (1) Invariant to equivalent utility representations; (2) Pareto optimality; (3) Independence of irrelevant alternatives; and (4) Symmetry. By solving the above optimization problem, we can obtain a fair profit distribution as follows:

$$\begin{aligned} \mathbb{U}_{ISP}^{S3} &= \mathbb{U}_{ISP}^{s} + \frac{\mathbb{U}_{total}^{S3} - \mathbb{U}_{ISP}^{s} - \mathbb{U}_{CP}^{s}}{2}, \\ \mathbb{U}_{CP}^{S3} &= \mathbb{U}_{CP}^{s} + \frac{\mathbb{U}_{total}^{S3} - \mathbb{U}_{ISP}^{s} - \mathbb{U}_{CP}^{s}}{2}. \end{aligned} \qquad (4.24)$$

Then, the profit that PCP coalition should transfer to ISP coalition is $\mathscr{R} = \mathbb{U}_{ISP}^{S3} - \mathbb{U}_{ISP}^{S3'} = \mathbb{U}_{CP}^{S3'} - \mathbb{U}_{CP}^{S3}$.

For different traffic profiles, we illustrate the improvement of $\mathbb{U}_{ISP}^{S3}$ and $\mathbb{U}_{CP}^{S3}$ in comparison with the values on the corresponding starting point (i.e., $(\mathbb{U}_{ISP}^{S2}, \mathbb{U}_{CP}^{S2})$ as we have analyzed in Sect. 4.1.4.3) in Fig. 4.10.

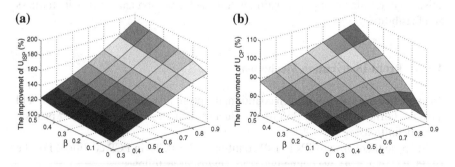

Fig. 4.10 The improvement of $\mathbb{U}_{ISP}^{S3}$ and $\mathbb{U}_{CP}^{S3}$ compared with the values on the corresponding starting point for different α and β

Fig. 4.11 An example of
Nash bargaining between ISP
and PCP coalitions $(\alpha, \beta) =$
$(0.6, 0.3)$

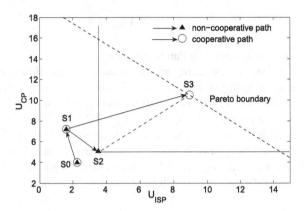

From Fig. 4.10, we can see that $\mathbb{U}_{ISP}$ increases by more than 110 %, and $\mathbb{U}_{CP}$ increases by more than 70 %, compared with the starting point.

Specifically, for $\alpha = 0.6$ and $\beta = 0.3$, the Nash bargaining between ISP and PCP coalitions is illustrated in Fig. 4.11, from which we can see that the corresponding starting point is $\left(\mathbb{U}_{ISP}^{S2}, \mathbb{U}_{CP}^{S2}\right) = (3.5180, 5.6450)$. According to Eq. (4.24), we can obtain $(\mathbb{U}_{ISP}^{S3}, \mathbb{U}_{CP}^{S3}) = (8.6508, 10.7778)$ as the final profit distribution. Then, the profit that PCP coalition should assign to ISP coalition is $\mathscr{R} = 3.7449$. Compared with the starting point, $\mathbb{U}_{ISP}$ increases by 145.9 %, and $\mathbb{U}_{CP}$ increases by 90.92 %. Thus, both ISP coalition and PCP coalition benefit much from this cooperation.

4.2.2 Profit Distribution Within Each Coalition

From the discussion in Sect. 4.2, we can see that PCP coalition should assign some profit $\mathscr{R}$ to ISP coalition in the cooperation. In this section, we will propose a mechanism to determine profit distribution within each coalition and discuss its fairness and feasibility.

4.2.2.1 Profit Distribution Mechanism

To guarantee the stability of each coalition, the profit distribution mechanism should have the fairness character. Before introducing such a mechanism, we first provide some definitions.

Suppose there are m ISPs in ISP coalition and n PCPs in PCP coalition. For the ith PCP $(1 \leq i \leq n)$, we define two *traffic matrices* as follows:

1. $\mathbf{T}_i = \left(t^i_{j,k}\right)_{m \times m}$, where $t^i_{j,k}$ denotes the amount of the ith PCP's traffic volume transmitted from the users in the jth ISP's network to the users in the kth ISP's network; and
2. $\widetilde{\mathbf{T}}_i = \mathrm{diag}(\tilde{t}^i_1, \tilde{t}^i_2, \dots, \tilde{t}^i_m)$, where $\tilde{t}^i_j$ denotes the amount of the ith PCP's traffic volume transmitted from its content servers to the users in the jth ISP's network (Note that this part of uploading traffic will be charged by the corresponding ISP on the ith PCP side).

According to the network model described in Sect. 4.1.1, the PCP traffic provided by P2P accounts for $1 - \beta$ proportion, and the other part is provided by servers of PCPs. Then, we have

$$\sum_{i=1}^{n} \left(\sum_{1 \le j,k \le m} t^i_{j,k} \right) = v_{\mathrm{p2p}} \cdot (1 - \beta) \quad \text{and} \quad \sum_{i=1}^{n} \left(\sum_{j=1}^{m} \tilde{t}^i_j \right) = v_{\mathrm{p2p}} \cdot \beta. \qquad (4.25)$$

Thus, in PCP coalition, the amount of traffic volume caused by the ith PCP accounts for a proportion

$$\varphi_i = \frac{\left(\sum_{1 \le j,k \le m} t^i_{j,k} \right) + \left(\sum_{j=1}^{m} \tilde{t}^i_j \right)}{v_{\mathrm{p2p}}}, \qquad (4.26)$$

From Eq. (4.25), it is clear that $\sum_{i=1}^{n} \varphi_i = 1$.

For ISP coalition, its two corresponding *aggregated traffic matrices* are defined by

$$\mathbb{T} = \sum_{i=1}^{n} \mathbf{T}_i, \quad \widetilde{\mathbb{T}} = \sum_{i=1}^{n} \widetilde{\mathbf{T}}_i.$$

Suppose $\mathbb{T} = \left(t_{j,k}\right)_{m \times m}$ and $\widetilde{\mathbb{T}} = \mathrm{diag}(\tilde{t}_1, \tilde{t}_2, \dots, \tilde{t}_m)$. Then, in the lth ISP's network (where $1 \le l \le m$), the amount of P2P traffic volume caused by PCP coalition on users' side is

$$\varpi_l = \left(\sum_{k=l}^{k=m} t_{l,k} \right) + \left(\sum_{k=l}^{k=m} t_{k,l} \right) + \tilde{t}_l. \qquad (4.27)$$

In addition, in the lth ISP's network, let v_l and b_l be the total traffic volume on users' side and the total bandwidth bought by all users with a flat price, respectively. It should be noted that $\sum_{l=1}^{m} b_l = b^{\mathrm{S0}}_{\mathrm{user}}$. Then, in the lth ISP's network, we can verify that the amount of the background C/S traffic volume is $v_l - \varpi_l$, and the free-riding P2P traffic volume is $v_l - b_l \cdot \xi_{\mathrm{user}}$ (where ξ_{user} is the bandwidth utilization rate assumed by ISP coalition when setting the flat price). According to the network model described in Sect. 4.1.1, clearly,

$$\sum_{l=1}^{m} [b_l \cdot \xi_{\text{user}} - (v_l - \varpi_l)] = v^{S0} \cdot \alpha.$$

In addition, we can deduce that

$$\sum_{l=1}^{m} (v_l - b_l \cdot \xi_{\text{user}}) = v_{\text{p2p}} \cdot (2 - \beta) - v^{S0} \cdot \alpha. \tag{4.28}$$

Thus, the lth ISP's contribution to the free riding of P2P traffic accounts for a proportion

$$\psi_l = \frac{v_l - b_l \cdot \xi_{\text{user}}}{v_{\text{p2p}} \cdot (2 - \beta) - v^{S0} \cdot \alpha}. \tag{4.29}$$

From Eq. (4.28), it is clear that $\sum_{l=1}^{m} \psi_l = 1$.

Consequently, we propose a fair and feasible profit distribution mechanism as follows. For a given $\mathscr{R}$, the profit that the ith PCP should assign to ISP coalition is $\mathscr{R} \cdot \varphi_i$ (where $1 \leq i \leq n$), and the profit that ISP coalition should assign to the lth ISP is $\mathscr{R} \cdot \psi_l$ (where $1 \leq l \leq m$). Consider the example in the previous section. Suppose $M_{\text{ISP}} = \{\text{ISP}_1, \text{ISP}_2, \text{ISP}_3\}$ and $M_{\text{PCP}} = \{\text{PCP}_1, \text{PCP}_2\}$. In addition, suppose

$$T_1 = \begin{pmatrix} 0.8255 & 1.6509 & 2.4764 \\ 1.6509 & 1.6509 & 3.3019 \\ 0.8255 & 1.6509 & 1.6509 \end{pmatrix},$$

$$T_2 = \begin{pmatrix} 2.4764 & 1.2382 & 3.7146 \\ 1.2382 & 2.4764 & 1.2382 \\ 4.9528 & 2.4764 & 3.7146 \end{pmatrix} \text{ and }$$

$$\widetilde{T}_1 = \text{diag}(1.4151, 2.1226, 3.1840),$$
$$\widetilde{T}_2 = \text{diag}(3.7146, 2.6533, 3.7146).$$

According to Eq. (4.25), we can verify that $v_{\text{p2p}}^{S3} = 56.0140$. Then, from Eq. (4.26), we can deduce that $(\varphi_1, \varphi_2) = (0.4, 0.6)$. Thus, the profits that PCP_1 and PCP_2 should assign to ISP coalition are $\mathscr{R} \cdot \varphi_1 = 1.4980$ and $\mathscr{R} \cdot \varphi_2 = 2.2469$. Then, for ISP coalition, suppose $(v_1, v_2, v_3) = (29.7479, 27.7250, 38.6127)$ and $(b_1, b_2, b_3) = (2.6349, 2.4322, 3.5469)$. Then, according to Eq. (4.29), we have $(\psi_1, \psi_2, \psi_3) = (0.3097, 0.2887, 0.4016)$. Thus, the profit that PCP coalition should assign to ISP_1, ISP_2, and ISP_3 are 1.1598, 1.0812 and 1.5040, respectively. Now, consider a more general example with $(\varphi_1, \varphi_2) = (0.4, 0.6)$. We suppose the numbers of users in ISP_1, ISP_2 and ISP_3 are N_1, N_2, and N_3, respectively, and the ratio of $N_1 : N_2 : N_3$ is $2 : 3 : 5$. Let $\mathbb{N} = N_1 + N_2 + N_3$. Moreover, suppose that in these networks, all the users have the same preferences and behaviors. Then, the initially bought bandwidth is proportional to the number of users, that is, $b_1 : b_2 : b_3 = N_1 : N_2 : N_3 = 2 : 3 : 5$.

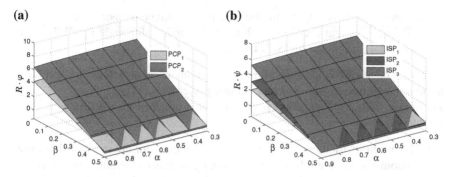

Fig. 4.12 a PCP and **b** ISP profit transfer with different α and β

The requirements of background traffic are also proportional to the number of users, i.e., $v_{CS1}^{S0} : v_{CS2}^{S0} : v_{CS3}^{S0} = N_1 : N_2 : N_3$. Besides, we assume P2P applications use a random peer selection scheme, and the contents are distributed uniformly among users. For the ith P2P application, suppose the average usage of each user is σ_i $(i = 1, 2)$. Then, we can obtain the traffic matrix of the ith PCP as following

$$\mathbf{T}_i = \sigma_i \cdot (1 - \beta) \cdot \begin{pmatrix} N_1 \cdot \frac{N_1}{N} & N_2 \cdot \frac{N_1}{N} & N_3 \cdot \frac{N_1}{N} \\ N_1 \cdot \frac{N_2}{N} & N_2 \cdot \frac{N_2}{N} & N_3 \cdot \frac{N_2}{N} \\ N_1 \cdot \frac{N_3}{N} & N_2 \cdot \frac{N_3}{N} & N_3 \cdot \frac{N_3}{N} \end{pmatrix} .$$

Moreover, based on our assumptions on users' behaviors and the above matrixes, we know that traffic provided by servers for each network is also proportional to $N_1 : N_2 : N_3$:

$$\tilde{\mathbf{T}}_i = \text{diag}(\sigma_i \cdot N_1 \cdot \beta, \ \sigma_i \cdot N_2 \cdot \beta, \ \sigma_i \cdot N_3 \cdot \beta).$$

Therefore, based on Eq. (4.29), we can get the ratio of three ISPs' contribution weight i.e. $\psi_1 : \psi_2 : \psi_3 = N_1 : N_2 : N_3 = 2 : 3 : 5$. Figure 4.12 illustrates the amount of profit transfer of each PCP and ISP in different traffic profiles ($0.3 \leq \alpha \leq 0.9$ and $0 \leq \beta \leq 0.5$). We can see that with the increase of α or β, such profit transfer decrease.

4.2.2.2 Discussion

The proposed mechanism guarantees the following two points,
Fairness. The fairness of this profit distribution mechanism is guaranteed by the following two characteristics:

(1) φ_i increases in accordance with the total traffic volume of the ith PCP;

(2) ψ_l increases accordingly with the total traffic volume on users' side in the lth ISP's network, but decreases with the total bandwidth bought by all the users with a flat price in the lth ISP's network.

It can be verified that this profit distribution mechanism has the following attributes: efficiency, symmetry, and dummy player [7, 8, 17].
Feasibility. This profit distribution mechanism is practical for the following three reasons:

(1) This profit distribution mechanism is compatible with the traditional Internet economic settlement. Transit ISPs do not join this cooperation, and thus, the transit traffic can still be charged according to the old economic agreements between transit ISPs and eyeball ISPs;
(2) All the information required by this profit distribution mechanism can be collected from ISPs and PCPs;
(3) The calculation of this profit distribution mechanism is easy.

4.3 Summary

Under the traditional Internet pricing mechanism, free-riding P2P traffic can cause an unbalanced profit distribution between PCPs and eyeball ISPs. This imbalance will drive eyeball ISPs to take actions against P2P and can finally impede the wide adoption of P2P technology. Therefore this chapter proposes a new cooperative profit-distribution model based on the concept of Nash bargaining, in which both eyeball ISPs and PCPs first form a coalition, respectively, and then cooperate to maximize their total profit. The fair profit distribution between the two coalitions is determined by Nash Bargaining Solution (NBS). To guarantee the stability of each coalition, a fair mechanism for profit distribution within each coalition has been designed. Such a cooperative profit-distribution method not only guarantees the fair profit distribution among network participators, but also improves the economic efficiency of the overall network system.

References

1. Cao, X.R., Shen, H.X., Milito, R., Wirth, P.: Internet pricing with a game theoretical approach:concepts and examples. IEEE/ACM Trans. Netw. **10**(2), 208–216 (2002)
2. Cocchi, R., Shenker, S., Estrin, D., Zhang, L.: Pricing in computer networks: motivation, formulation, and example. IEEE/ACM Trans. Netw. **1**(6), 614–627 (1993)
3. Debreu, G.: A social equilibrium existence theorem. Proc. Natl. Acad. Sci. USA. **38**(10), 886–893 (1952)
4. Dhamdere, A., Dovrolis, C.: Can ISPs be profitable without violating "network neutrality"? In: Proceedings of the International Workshop on Economics of Networked Systems 2008, pp. 13–18, ACM (2008)

5. Dimitropoulos, X., Hurley, P., Stoecklin, A.K.M.: On the 95-percentile billing method. In: Proceedngs of PAM 2009, Seoul, South Korea (2009)
6. He, H., Xu, K., Liu, Y.: Internet resource pricing models, mechanisms, and methods. Netw. Sci. 1, 48–66 (2012)
7. Ma, R.T.B., Chiu, D., Lui, J.C.S., Misra, V., Rubenstein, D.: Internet economics: the use of shapley value for ISP settlement. IEEE/ACM Trans. Netw. 18(3), 775–789 (2010)
8. Ma, R.T.B., Misra, V., Chiu, D., Rubenstein, D., Lui, J.C.S.: On cooperative settlement between content, transit and eyeball internet service providers. IEEE/ACM Trans. Netw. 19(3), 802–815 (2011)
9. MacKie-Mason, J.K., Varian, H.R.: Pricing congestible network resources. IEEE J. Sel. Areas. Commun. 13(7), 1141–1149 (1995)
10. Moulin, H.: Axioms of cooperative decision making. Cambridge University Press, Cambridge (1988)
11. Nash, J.F.: The bargaining problem. Econometrica 28, 155–162 (1950)
12. Norton, W.B.: Video internet: the next wave of massive disruption to the U.S. peering ecosystem. In: Equinix white papers (2007)
13. Rasmusen, E.: Games and information: an introduction to game theory, 4th edn. Blackwell Publishing, Oxford (2007)
14. Rodriguez, P., Tan, S.M., Gkantsidis, C.: On the feasibility of commercial, legal P2P content distribution. ACM SIGCOMM Comput. Commun. Rev. 36(1), 75–78 (2006)
15. Valancius, V., Lumezanu, C., Feamster, N., Johari, R., Vazirani, V.V.: How many tiers? Pricing in the Internet transit market. In: Proceedings of the SIGCOMM 2011, pp. 194–205, ACM (2011)
16. Wang, Q., Chiu, D., Lui, J.C.: ISP uplink pricing in a competitive market. In: Proceedings of International Conference on Telecommunications 2008, pp. 1–6, IEEE (2008)
17. Winter., E.: The shapley value. In: Aumann, R.J., Hart, S. (eds.) The Handbook of Game Theory. North-Holland, Amsterdam (2002)
18. Xie, H., Yang, Y.R., Krishnamurthy, A., Liu, Y.G., Silberschatz, A.: P4P: provider portal for applications. In: Proceedings of SIGCOMM 2008, pp. 351–362, ACM (2008)
19. Yaiche, H., Mazumdar, R.R., Rosenberg, C.: A game theoretic framework for bandwidth allocation and pricing in broadband networks. IEEE/ACM Trans. Netw. 8(5), 667–678 (2000)

Chapter 5
Pricing in Multi-Interface Wireless Communication Markets

This chapter focuses on the dynamic game relationship in the Internet service market, where ISPs provide services and multi-interface mobile users select services. Through the modeling of the Internet service market, service composition and the users, the bargaining in the exclusive monopoly market and the dynamic game procedure in the oligopoly market were analyzed by using the non-cooperative game theory. In addition, it was proved that under the idealized condition, the ISP could gain more profits if it offers various unique service combinations. It shows that the subgame perfect Nash equilibrium of the finitely repeated game is that at every stage, all of the ISPs adjusts to make their strategies the same, and the final result is that all of the ISPs develop a service combination for each user. For the infinitely repeated game under certain circumstances, there is a specific subgame perfect Nash equilibrium, which is that all ISPs don't adjust about their strategies. At last, the suggestions to ISPs on the pricing process were given by the confirmation of the ISP price monopoly position in the Exclusive Monopoly Market through experiments, and the methods of deciding service combinations according to the features of user groups will be developed by testing the ISP dynamic game procedure in the Oligopoly Market.

5.1 Background

Mobile Internet has become an important trend in the development of networks. As shown in Fig. 5.1, mobile hosts (MH) using network services are usually equipped with several network interfaces (NI), and can access the Internet through different ISPs. For example, an iPhone can access the Internet via GPRS, 3G, WiFi, etc. With the fierce competition among ISPs for the market and profits, users also have the right to choose network services and decide the way to use them. From the economic perspective, there is a dynamic game existing between ISPs and users [5, 11, 14, 15].

K. Xu et al., *Internet Resource Pricing Models*,
SpringerBriefs in Computer Science, DOI: 10.1007/978-1-4614-8409-7_5,
© The Author(s) 2014

Fig. 5.1 Mobile hosts access
the Internet with multiple
network interfaces

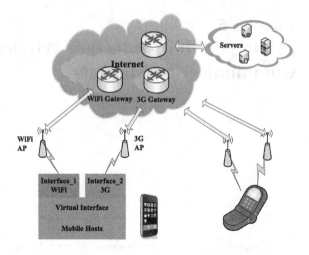

In economics, according to the competition type, the market is divided into the perfect competitive market [18] and the imperfect competitive market [8]. The former is the ideal condition. The concept of imperfect competition, proposed by J.M. Clark, refers to the fact that perfect competition does not actually exist [23]. In the imperfect competitive market, monopoly, to some degree, exists and is divided into the exclusive monopoly market [2], the oligopoly market [7] and the joint monopoly market [17], according to their organization types.

This chapter describes the research on the dynamic game relationship between ISPs and users by using the non-cooperative game theory in economics, because the participants in competition can hardly reach a binding proposal [19].

5.2 Modeling

This section introduces three types of Internet service markets (the exclusive monopoly market, the oligopoly market and the joint monopoly market), builds a general service composition model by investigating ISPs' service composition, and makes a multi-interface mobile host user demand model according to different market styles.

We first explain marks and variables used in this chapter. If not stated, the meaning of marks is determined according to Table 5.1.

Table 5.1 Notations

Notations	Description	Reference value
USERS	Number of the multi-interface mobile hosts	10^5
I	Number of ISPs, i is the index	[1, 10]
α	ISP cost parameter related to the volume of services and user scale	0.01
β	The part of ISP cost with no relationship with service usage	10^5
λ	ISP evaluation factor	1, 2
ρ	Tradeoff parameter for user fees and performance	$[0, \infty)$
$Plan_i$	The matrix that describes the service combination provided by ISP_i	–
π	Profit of ISP_i	–
$Income_i$	Total income of ISP_i	–
$Cost_i$	Total cost of ISP_i	–

5.2.1 Internet Service Market

There are two groups in the Internet service market, ISPs and multi-interface mobile users. This section mainly focuses on the three types of the imperfect competition market of the Internet service market. In different types of markets, the game relationship between ISPs and users is fairly various [9].

5.2.1.1 Exclusive Monopoly Market

Exclusive monopoly market refers to the market in which one company has the exclusive control over both the production and the operation of the whole business [2]. To the Internet service market, in the exclusive monopoly market, only one ISP provides service for mobile host users. From the economic perspective, this ISP can make and control the service price, achieving the maximum profits.

In the exclusive monopoly market, the game relationship between the ISP and multi-interface mobile host users lies in the fact that while the ISP offers services and regulates the price, users choose to buy services provided according to their own demands (relevant to the service price).

5.2.1.2 Oligopoly Monopoly Market

Oligopoly monopoly market refers to a market where several companies produce and sell some certain products or services, each of which has a certain client share under competition [7]. This market type in the marketing economy is common in business like petroleum, living goods and the Internet service market.

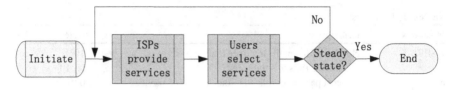

Fig. 5.2 Dynamic game procedure

In the oligopoly monopoly market, several ISPs provide different kinds of service composition, and users can choose more than one ISP and service composition based on their need, which forms a dynamic game between ISPs and users. As shown in Fig. 5.2, ISPs regulate service composition from time to time to achieve the maximum client group and profits, and then users re-choose ISPs and service composition. This process is repeated until the market is finally in a stable state, and the game is then over.

5.2.1.3 Joint Monopoly Market

Joint monopoly market refers to a market where two or more companies, by means of restrictive practice, or solidary behaviors, jointly control the production and sales of a business. The joint monopoly market includes many types, including the temporary price agreement, Cartel, Syndicate, Trust, Konzem, etc. [17].

Since this kind of monopoly impedes market competition and affects social development, it is sanctioned worldwide by law.

For the Internet service market, this type is greatly controlled by man. Its price strategy is made in agreement by companies, enabling companies to get stable profits [21, 22, 24]. In terms of games, this kind of monopoly is unstable. This joint monopoly market will be broken when a certain ISP adopts new techniques and regulates service composition so as to cut costs and get more users and profits. Then, the market will turn into an oligopoly market.

5.2.2 ISPs' Service Composition Model

By researching the major ISPs in the market in China, we can get: first, ISPs can provide services including voice call, SMS, data traffic and video call; second, ISPs can set different service intervals to meet user demands. Basically, clients need to pay the prepayment, which is not included into the free part. Therefore, the service has three parts: prepayment, free interval service usage, and charging standards when the amount of service goes beyond the free interval. It is an interval pricing method in price discrimination [3], aiming to gain profits from service that clients fail to use. To be more specific, we have ISPs' service composition model. Assuming an ISP can provide n kinds of service, the number of the k-th service interval provided by ISP_i is m_{ki}, $m_i = \sum_{k=1}^{n} m_{ki}$, $(k = 1, 2, \ldots, n)$. Then, the service composition model of ISP_i is:

$$Plan_i = \begin{pmatrix} x_{11} & x_{12} & x_{13} \\ \vdots & \vdots & \vdots \\ x_{m_i 1} & x_{m_i 2} & x_{m_i 3} \end{pmatrix}, \quad (i = 1, 2, \ldots, n)$$

where the first line of the matrix refers to the prepayment of a certain service, the second the free interval service usage, and the third the charging standards when the amount of service goes beyond the free interval.

5.2.3 Multi-Interface Mobile Host User Model

By surveying multi-interface mobile users who use ISPs' service, we find that the service demand of user group is subject to certain statistical regularity (e.g. normal distribution). We use *Demand* to denote the service demand and $demand(x)$ to denote its density function, both of which can be got by data fitting. Based on different market types, we use two service demand models to illustrate user service demands.

5.2.3.1 Constant Elasticity Demand (CED) Model

In the exclusive monopoly market, there is only one ISP. Then, users have no choice but to use services provided by the ISP. Under such circumstances, users' average service demand is related to the changes of the service price. This demand is called the fixed elastic demand, built on the alpha-fair model [16]. Here, the function relationship between the average demand and the price is:

$$\bar{Q}(price) = \left(\frac{val}{price} \right)^{\lambda}$$

where val refers to the ISP's evaluation factor, $price$ the service price set by the ISP, and λ the sensitivity of users' demand to the changes of the price [20].

5.2.3.2 Optimal Utility Demand Model

In the oligopoly market and the joint monopoly market, multi-interface mobile hosts can access different ISP networks, and then users have more than one choice. An optimal utility demand model means that users always choose ISPs and service composition to maximize their own utilities.

Users' utility function refers to the optimization of the cost and performance. We choose the dual objective function with the optimal cost and performance:

$$\min \ Z = Payment + \rho Perf Measurement, \ (\rho \in [0, \infty))$$

where ρ is the compromise parameter of the cost and performance, determined by the demand of mobile hosts. The smaller value ρ has, the lower cost users prefer, which proves that users give importance to the performance.

5.3 Analysis of the Dynamic Game Process

This section first computes the payment of multiple interface mobile hosts and the profits of the ISPs, and then analyzes the dynamic relationship between users and ISPs in the two markets.

5.3.1 Payment Computing of Multi-Interface Mobile Hosts

Mobile hosts can access different ISPs through multiple interfaces, and then choose more than one ISP and service composition. From the service composition, we know that for a certain user, if he/she chooses ISP_i's service interval $p(x_{p1}, x_{p2}, x_{p3})$, with service usage being q, the payment is:

$$Payment(q, i, p) = \begin{cases} x_{p1}, & q \leq x_{p2} \\ x_{p1} + (q - x_{p2}) \times x_{p3}, & q > x_{p2} \end{cases}$$

If more than one service is chosen, all payments will add up.

5.3.2 Computing of ISP Profits

The profits of ISPs refer to the difference between the total income and the total cost. For example, the profit of ISP_i is:

$$\pi_i(Q) = Income_i(Q) - Cost_i(Q)$$

An ISP's total income and total cost will be computed as follows.

5.3.2.1 Total Income Computing

We assume that at t, the total number of users is $u(t)$, the number of users possessed by ISP_i is $u_i(t)$, and the number of users choosing ISP_i's interval k is $u_{ik}(t)$. Since users can choose more than one ISP and service composition, then:

$$\sum_{k=1}^{n} u_{ik}(t) \geq u_i(t), \quad \sum_{i=1}^{I} u_i(t) \geq u(t), \quad (i = 1, 2, \ldots, I)$$

From the user demand model, we know that the demand is in line with certain statistical regularity and its density function is $demand(x)$. The method of sharding is used to compute payments. We divide service demands into several parts, compute the payment in each part and add them up. Details are shown as follows:

Step 1: We divide users' demands into s parts, and compute the number of users in each part:

$$users = u(t) \times P(ks \leq X < (k+1)s)$$
$$= u(t) \times \int_{ks}^{(k+1)s} demand(x)dx, \quad (k = 0, 1, \ldots)$$

End condition: $users \leq \varepsilon$ (ε is a boundary threshold constant).

Step 2: Every user chooses one certain kind of service composition offered by one ISP according to his/her utility maximum principle. We assume the service demand composition of a certain user is q_j, and the service composition of ISP_i chosen by the user is $p_m | m = 1, 2, \ldots, m_i$. For the k-th, the gross income of ISP_i (cost included) is:

$$Income_i^k = \sum_{j=1}^{users} \sum_{m=1}^{m_i} Payment(q_j, i, p_m)$$

Step 3: The total income of ISP_i is the sum of k parts:

$$Income_i = \sum_k Income_i^k$$

5.3.2.2 Cost Computing

Cost here does not include investment in fixed assets. The scenario in this chapter is different from that in [20]. Since users use mobile hosts, the distance changes dynamically. Therefore, it is impossible to estimate the relationship between the cost and the distance. However, the distance in this problem is a random variable, and then we can ignore its macroscopic impact on the cost. Lin et al. [13] present that the relationship between the average cost on equipments per month and network scale is logarithmic. In addition, ISPs' cost here is highly relevant to service usage. Then, we assume ISPs' cost function can be expressed as follows:

$$Cost(q) = \alpha \log(u_i(t) * q) + \beta$$

where α is the relationship parameter between service usage and the user scale, and β is the part not related to service usage in ISP's cost.

5.3.3 Analysis of the Dynamic Game Process in Two Types of Markets

5.3.3.1 Exclusive Monopoly Market

In the exclusive monopoly market, ISPs can absolutely take control of the market price, but users decide the amount of service to buy. Then, the relationship between ISPs and users belongs to the bargaining game [4].

The process of the game between ISPs and users is shown in Fig. 5.3. At first, ISPs provide service p_0 and the service usage is q_0. Then, the payment is $(-Payment(q_0, 1, p_0), \pi(q_0))$. When ISPs put service p_1 into the market to increase the price, service usage decreases by Δq_1. Then the payment at this time is $(-Payment(q_0 - \Delta q_1, 1, p_1), \pi(q_0 - \Delta q_1))$. When ISPs make the price lower, the service usage increases by Δq_2. The payment here is $(-Payment(q_0 + \Delta q_2, 1, p_2), \pi(q_0 + \Delta q_2))$. If ISPs are satisfied with their profits, the bargaining game is over; if not, the game will continue.

Fig. 5.3 Bargain process between ISPs and users

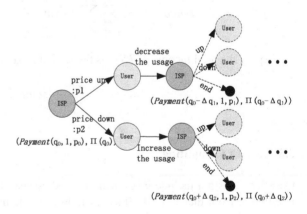

5.3.3.2 Oligopoly Market

We know that the joint monopoly market will turn into the oligopoly market once the price alliance is broken. As we focus on the process of dynamic games, we would like to analyze both these two markets.

In the oligopoly market, there are many ISPs and a variety of service composition for users to choose from. Participants in the game are ISPs, whose strategic space here is {adjustment, non-adjustment}, and the payment function is the profit of the ISP. For example, we assume that there are two service providers, ISP_1 and ISP_2, providing the same service composition at first with their profits being (5, 5). If only one of them adjusts the service composition, his/her profit will rise to 6 while the other's will drop to 2. If both of them adjust the service composition, they will share the market again with their profits being (3, 3).

We assume ISP_1 first decides to adjust, and then ISP_2 observes the decision made by ISP_1, the game process of which is illustrated in Fig. 5.4. We denote adjustment by Y and non-adjustment by N. ISP_1's strategic space is {Y, N}. ISP_2 has four pure strategies, which are: (1) no matter whether ISP_1 adjusts or not, ISP_2 will adjust; (2) ISP_2 follows ISP_1's decision; (3) if ISP_1 adjusts, ISP_2 will not adjust, and vice versa; (4) no matter whether ISP_1 adjusts or not, ISP_2 will not adjust. Therefore, ISP_2's strategic space is {{YY}, {YN}, {NY}, {NN}}. The strategy in the game is expressed in Table 5.2.

From the analysis, we find that there is only one pure strategy Nash equilibrium in this game, which is {Y, {YY}}. As {YY} can reach the Nash equilibrium in both subgame (a) and subgame (b), {Y, {YY}} is the only subgame perfect Nash equilibrium in this game.

However, this problem cannot be solved by analyzing only one game. After both sides adjust their service composition, they launch another game, resulting in a game repetition. If it is a finitely repeated game, the only subgame perfect Nash equilibrium is to choose {Y, {YY}} in each stage of the game. If it is an infinitely repeated game,

Fig. 5.4 Game tree of the two ISPs

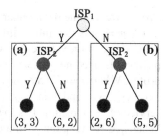

Table 5.2 Two ISPs' strategic form

ISP_1	ISP_2			
	YY	YN	NY	NN
Y	(3, 3)	(3, 3)	(6, 2)	(6, 2)
N	(2, 6)	(5, 5)	(2, 6)	(5, 5)

when ISP_1 and ISP_2 choose not to adjust, the profits of both sides are higher than when choosing to adjust. Under the latter condition, ISPs can choose the grim strategy: (1) choosing not to adjust at first; (2) choosing not to adjust until one side starts to adjust, and then always choosing the adjustment strategy. According to the folk theorem, discount factor $\delta^* < 1$ exists. Then when $\delta \geq \delta^*$, the strategy of not to adjust is a specific subgame perfect Nash equilibrium. We compute the δ^* as follows:

$$6 + 3\delta + 3\delta^2 + \cdots \leq 5 + 5\delta + 5\delta^2 + \cdots$$

or:

$$6 + \frac{3\delta}{1 - \delta} \leq \frac{5}{1 - \delta}$$

Thereafter, we get $\delta \geq \frac{1}{3}$ and $\delta^* = \frac{1}{3}$. Here, both ISP_1 and ISP_2 can gain higher profits if they do not choose to adjust.

Hence, we draw the conclusion that the process of dynamic repeated games between ISPs follows the following rules: first, when emerging ISPs adopt the same strategies as original ISPs, users will re-choose ISPs and service composition; second, when some ISPs provide new service composition to attract more clients and gain higher profits, users choose the one with the maximal utility; third, after ISPs choosing non-adjustment suffer client and profit declination, they will also adjust their composition strategies. The second and the third steps happen in turn until the market reaches a temporarily stable state.

We assume there are I ISPs with G stage games, and $G(T)$ refers to the repeated game after repeating T times. We hence analyze $G(T)$ in the way shown as follows when it is a finite repeated game ($T < \infty$) and an infinite game ($T = \infty$) respectively.

Theorem 1 *Under ideal conditions, a certain ISP can get more users and higher profits after offering service composition different from other ISPs'. Game G between ISPs and users has the only Nash equilibrium. That is to say, all ISPs adjust to the same service composition, sharing users and profits.*

Proof We assume the number of ISPs is I and the number of services is $n(n \geq 0)$.

(1) $n = 0$ is the initial situation where none of the ISPs has provided any service composition yet, i.e., $(0, 0, x_{03})$. The users choose ISPs randomly, since the payment is fixed no matter which ISP they choose. Therefore, ISPs share users and profits. When $n = 1$, ISP_j provides service composition 1, which is represented by (x_{11}, x_{12}, x_{13}), and then $x_{11}/x_{12} < x_{03}$, which means that the average price per unit service of composition 1 is lower than that when there is no service composition. When $q \in [0, x_{12}]$, the payment is $Payment(q, j, 1) = x_{11}$ if ISP_j's service composition 1 is chosen; when $q \in [x_{12}, \infty)$, $Payment(q, j, 1) = x_{11} + (q - x_{12}) \times x_{13}$; when service composition 1 is not chosen, $Payment(q, i, 1) = q \times x_{03}$. Then, we deduce that when $x_{13} \leq x_{03}$ and $q \in \left(\frac{x_{11}}{x_{03}}, \infty\right)$, or when $x_{13} > x_{03}$ and $q \in \left(\frac{x_{11}}{x_{03}}, \frac{x_{12} \times x_{13} - x_{11}}{x_{13} \times x_{03}}\right]$, users prefer service composition 1. Consequently, ISP_j can

gain more users and profits. Then, other ISPs will also provide service composition 1, and share the profits. Thus, the proposition is true.

(2) When $n = k$, after several rounds of games, there are I ISPs providing k kinds of service compositions. We assume this proposition is true.

(3) When $n = k + 1$, if ISP_j provides service t', which is represented by $(x_{t'1}, x_{t'2}, x_{t'3})$, the $(k + 1)$-th kind, which is between t and $t + 1$, we get $x_{t1} < x_{t'1} < x_{(t+1)1}, x_{t2} < x_{t'2} < x_{(t+1)2}, t \in \{1, 2, \ldots, k\}$. Since services provided by ISPs are less expensive than normal data flow, then:

$$x_{(t+1)1} - x_{t1} < x_{t3} \times (x_{(t+1)2} - x_{t2}) \tag{5.1}$$

$$x_{t'1} - x_{t1} < x_{t3} \times (x_{t'2} - x_{t2}) \tag{5.2}$$

$$x_{(t+1)1} - x_{t'1} < x_{t'3} \times (x_{(t+1)2} - x_{t'2}) \tag{5.3}$$

For users whose service usage is $q \in [x_{t2}, x_{(t+1)2}]$, when they choose service interval t, $Payment(q, i, t) = x_{t1} + (q - x_{t2}) \times x_{t3}$; when they choose $(t + 1)$, $Payment(q, i, t + 1) = x_{(t+1)1}$. Then, when $q \in \left[x_{t2}, x_{t2} + \frac{x_{(t+1)1} - x_{t1}}{x_{t3}} \right]$, users prefer to choose t; when $q \in \left[x_{t2} + \frac{x_{(t+1)1} - x_{t1}}{x_{t3}}, x_{(t+1)2} \right]$, users prefer $(t + 1)$. In the following part, we will discuss users' choice when service interval is t'.

For user group $q \in [x_{t2}, x_{t'2}]$, if service interval t' is chosen, $Payment(q, j, t') = x_{t'1}$. Hence, we know when $Payment(q, j, t') < Payment(q, i, t)$ and $Payment(q, j, t') < Payment(q, i, t + 1)$, users prefer t'. Then, the simultaneous equations are as follows:

$$\begin{cases} x_{t'1} < x_{t1} + (q - x_{t2}) \times x_{t3} \\ x_{t'1} < x_{(t+1)1} \\ q \in [x_{t2}, x_{t'2}) \end{cases} \tag{5.4}$$

Through (5.2) and (5.4), we know that users, whose service usage is $q \in \left(\frac{x_{t'1} - x_{t1}}{x_{t3}} + x_{t2}, t_{t'2} \right)$, will get the lowest payment if choosing t'.

For users whose service usage is $q \in [x_{t'2}, x_{(t+1)2}]$, if they choose t', we have $Payment(q, j, t') = x_{t'1} + (q - x_{t'2} \times x_{t'3})$. Accordingly, the simultaneous equations are:

$$\begin{cases} x_{t'1} + (q - x_{t'2}) \times x_{t'3} < x_{t1} + (q - x_{t2}) \times x_{t3} \\ x_{t'1} + (q - x_{t'2}) \times x_{t'3} < x_{(t+1)1} \\ q \in [x_{t'2}, x_{(t+1)2}] \end{cases} \tag{5.5}$$

By deducing (5.5), we find that when service usage q can meet the following conditions, users pay the least if choosing t'.

$$q \times (x_{t'3} - x_{t3}) + (x_{t'1} - x_{t1}) - (x_{t'2} \times x_{t'3} - x_{t2} \times x_{t3}) < 0 \tag{5.6}$$

$$x_{t'2} \leq \frac{x_{(t+1)1} - x_{t'1}}{x_{t'3}} + x_{t'2} \tag{5.7}$$

The right side of (5.7) is denoted by M. When $x_{t'3} \leq x_{t3}$, we can prove that inequality (5.6) is always true, and then $x_{t'2} \leq q < M$. When $x_{t'3} > x_{t3}$, $q < \frac{(x_{t'2} \times x_{t'3} - x_{t2} \times x_{t3}) - (x_{t'1} - x_{t1})}{x_{t'3} - x_{t3}}$ (the expression on the right side is denoted by N) and $x_{t'2} \leq q < \frac{x_{(t+1)1} - x_{t'1}}{x_{t'3}} + x_{t'2}$, we can deduce $N > x_{t'2}$. By computing, when $x_{t3} < x_{t'3} < \frac{x_{t3} \times (x_{(t+1)1} - x_{t'1})}{x_{(t+1)1} - x_{t1} - x_{t3} \times (x_{t'2} - x_{t2})}$ (the expression on the right side is denoted by R; we can prove that $R > x_{t3}$), $N > M$, accordingly, $x_{t'2} \leq q < M$; when $R < x_{t3}$, $N < M$, accordingly, $x_{t'2} \leq q < N$.

To sum up, when $x_{t'3} < R$ and $q \in \left(\frac{x_{t'1} - x_{t1}}{x_{t3}} + x_{t2}, M \right)$, or when $x_{t'3} > R$ and $q \in \left(\frac{x_{t'1} - x_{t1}}{x_{t3}} + x_{t2}, N \right)$, users pay the least if choosing t'. Hence, the newly provided service interval brings ISP_j more users and profits. Then, other ISPs will imitate ISP_j to provide the same service, sharing the market again.

Hence, this proposition is proved.

From *Theorem 1*, we know stage game G has the only Nash equilibrium (all ISPs adjust their strategies). When $G(T)$ is a finitely repeated game, $G(T)$'s subgame perfect Nash equilibrium is to take the only Nash equilibrium in each stage of G [19]. When $G(T)$ is an infinitely repeated game and $\delta^* < 1$, for all $\delta \geq \delta^*$, that all of the ISPs choose non-adjustment is a specific subgame perfect Nash equilibrium solution to $G(T)$.

If users' service demand is q and service p of ISP_i is chosen, users' unit service fee is:

$$\bar{c}(q) = \frac{Payment(q, i, p)}{q}$$

Theorem 2 *The unit service fee should be greater than ISPs' unit service cost. Otherwise, they will lose their money.*

Proof When the service usage $q \leq x_{p2}$ and $\bar{c}(q) = \frac{x_{p1}}{q}$, we get the derivate $\frac{d\bar{c}(q)}{dq} = -\frac{1}{q^2}$, whose minimum occurs when $q = x_{p2}$. When $q > x_{p2}$ and $\bar{c}(q) = \frac{x_{p1} + (q - x_{p2} * x_{p3})}{q}$, we get the derivate $\frac{d\bar{c}(q)}{dq} = -\frac{-x_{p1} + x_{p2} * x_{p3}}{q^2} > 0$, whose minimum also occurs when $q = x_{p2}$. To sum up,

$$min(\bar{c}(q)) = \frac{x_{p1}}{q} \bigg|_{q=x_{p2}} = \frac{x_{p1}}{x_{p2}}$$

In the real world, games are usually finite, so we can take the finite repeated game into consideration. From *Theorem 1*, multiple ISPs driven by the market will be the finite repeated game until every user gains the most appropriate service composition. From *Theorem 2*, when users all have their own service composition, as $q \leq x_{p2}$, the unit service fee is the lowest and closest to the unit service cost. Therefore, the pure profit of ISPs will decrease. By summing up the above theorems and analyses, we can get the following conclusions:

Conclusion 1: In the oligopoly market, ISPs need to provide multiple service compositions if they would like to acquire higher profits. Competition usually results in the situation where ISPs share users and the market, and the total profit will decline in accordance with the increase of games.

Conclusion 2: Under the ideal condition, the result of repeated games is that ISPs provide a specific service composition for each user.

Conclusion 3: When every user has an appropriate service composition, the unit service fee is the lowest and closest to the unit service cost.

In April 2013, China Telecom, one of the largest mobile service providers in China, starts to provide the building block service compositions [6], allowing users to choose highly customized service compositions and pushing the market to the ideal condition mentioned in Conclusion 2.

5.4 Simulation and Analysis

5.4.1 Survey on User Groups

As data traffic service is more typical, we choose it to conduct a survey and analysis of user groups. This method could also be used to other Internet services.

There are two ways here. The first one is to collect the half-year detailed records of 20 students' mobile service usage (from September 2011 to February 2012) in the lab. This method features a small number of users who belong to a single group. The second one is to collect the data traffic information of network users through online questionnaires (questionnaire contents include preferred ISPs, data traffic plans and the data traffic from November 2011 to February 2012). We get 792 effective questionnaires in total. This method is widely used and features a large number of users who belong to diverse groups.

After analyzing those detailed records of students, its flow distribution obeys the bell-shaped distribution. We speculate the reason is that students within a group share many similarities, such as the consumption level and the living environment. Then, data collected are very likely to follow similar regularities (average and variance).

We use normal probability paper to test all data because it is relatively direct and easier to confirm whether the overall data are normally distributed. If they are normally distributed, they are of linear distributed [10]. In Matlab, we can conduct the test with one sample normal distribution LillieforsTest [12]. If the test result is zero, the data are in accordance with the normal distribution.

We adopt LillieforsTest to test user data. If its result is nullo, the data obeys the normal distribution. We can get the average and the variance of user data through normal distribution fitting. Figure 5.5 shows the fitting of network users' two-month

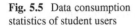

Fig. 5.5 Data consumption
statistics of student users

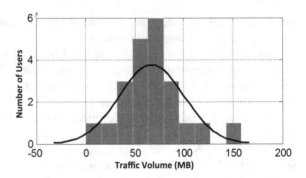

Fig. 5.6 Data consumption
statistics of network users

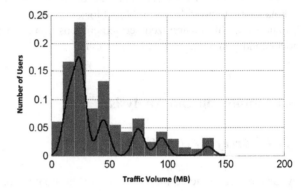

statistical data as well as the test result by using normal probability paper. Figure 5.6
is the fitting of network users' two-month statistical data.

From the data traffic accumulated during the four months, there are many peak
points, which are not normally distributed. It is possible because the survey is widely
extended. As different groups have different consumption levels and demands, their
usage of data traffic is also very different.

From the above analysis, we know that different user groups have different
demands for data traffic. But within the same group, their demands are to some
degree similar. Therefore, it is of great significance for ISPs to provide different
service compositions for different groups.

In our work, we can first choose to research the regularity of two different groups,
i.e., the overlap of two normal distribution density functions with different averages
and variances. This method can also be used to research multiple different groups.

We assume there are two groups in normal distribution whose density functions are
denoted by $f_1(x, \mu_1, \sigma_1)$ and $f_2(x, \mu_2, \sigma_2)$ respectively, where $\mu_1 = 30$, $\sigma_1 = 8$,
$\mu_2 = 70$ and $\sigma_2 = 11$. Assume these two groups are in the same quantitative
proportion, and then the probability density of their accumulated sum is:

$$demand(x) = \frac{1}{2}(f_1(x, \mu_1, \sigma_1) + f_2(x, \mu_2, \sigma_2))$$

$$= \frac{1}{2}\left(\frac{1}{\sigma_1\sqrt{2\pi}}e^{\frac{(x-\mu_1)^2}{2\sigma_1^2}} + \frac{1}{\sigma_2\sqrt{2\pi}}e^{\frac{(x-\mu_2)^2}{2\sigma_2^2}}\right)$$

We apply the function above to the total income computing of ISPs, and compute the population in k:

$$users = \text{USER} \times \int_{ks}^{(k+1)s} demand(x)dx$$

$$= \frac{1}{2}\text{USER} \times \left(\int_{ks}^{(k+1)s} f_1(x, \mu_1, \sigma_1)dx + \int_{ks}^{(k+1)s} f_1(x, \mu_1, \sigma_1)dx\right)$$

$$= \frac{1}{2}\text{USER} \times (F_1((k+1)s) - F_1(ks) + F_2((k+1)s) - F_2(ks))$$

$$\left(Y = \frac{X - \mu}{\sigma}\right) \text{ to standard normal distribution}$$

$$= \frac{1}{2}\text{USER} \times \left(\Phi\left(\frac{(k+1)s - \mu_1}{\sigma_1}\right) - \Phi\left(\frac{ks - \mu_1}{\sigma_1}\right)\right.$$

$$\left.+\Phi\left(\frac{(k+1)s - \mu_2}{\sigma_2}\right) - \Phi\left(\frac{ks - \mu_2}{\sigma_2}\right)\right)$$

where $\Phi(x)$ is the cumulative distribution function of standard normal distribution.

5.4.2 Analysis of the Exclusive Monopoly Market

This section first describes the influence of price on the average demand, and then the influence of ISPs' price on their profits and service usage.

5.4.2.1 Relationship Between Users' Average Demand and Price

In the exclusive market, the total demands of user groups obey the normal distribution. Users' average demand is in accordance with the CED model. In the test, we define $val_1 = 1$, $val_2 = 2$, $\lambda_1 = 1.5$ and $\lambda_2 = 3$. The relationship between the average demand of users and the price is shown in Fig. 5.7.

From Fig. 5.7, we find that when val is specified, the bigger λ is, the more sensitive users' average demand for the price is.

Fig. 5.7 Relation between
the average demand of users
and the price

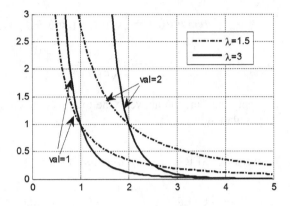

Fig. 5.8 Relationship
between the amount of
service consumed and
the price

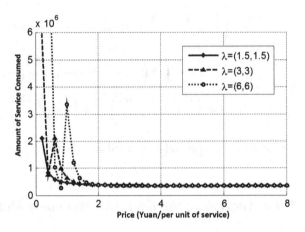

Fig. 5.9 Relationship
between the revenue of ISP
and the price

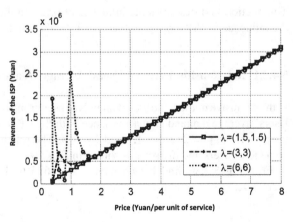

5.4.2.2 Pricing by ISPs

Simulation scene: we assume there is one ISP with no service composition, i.e. $(0, 0, x_{03})$, the number of users is denoted by $USER$, the val of the two kinds of user groups are assigned to 1 and 2 respectively, and the λ is assigned to (1.5, 1.5), (3, 3) and (6, 6).

With specified data, we find out how the profit of the ISP and the amount of service used vary with the price, which is shown in Figs. 5.8 and 5.9.

Test analysis: When $price < 0.5$, though the amount of service used is large, the profit is not high since the price is low; when $0.5 < price < 1.5$, the amount of service used is relatively large, and there is an increase in the profit; when $price > 1.5$, the amount of service used decreases significantly, but the profit keeps increasing as the price is high. Therefore, ISPs can reach a balance between the profit and the amount of service used when $0.5 < price < 1.5$.

5.4.2.3 Summary of the Exclusive Monopoly Market

From the analyses above, we know that, in the exclusive monopoly market, if the ISPs blindly raise the service price, although they will gain huge profits, the consumption of their service will decrease. In this market, ISPs should make low, user-oriented price if they really want to serve the public. If they tend to make their services upscale, a premium price should be made. But if they want both, they need to make the price medium.

5.4.3 Analysis of the Oligopoly Market

Users can have multiple choices when many ISPs exist. Then users will always choose the best ISPs and service compositions. Under this condition, ISPs compete to get much more share of users and profits. This section analyzes this problem respectively in aspects of ISPs entering the market and making service compositions.

5.4.3.1 Multiple ISPs Entering the Market

Laboratory simulation scene: the number of ISPs increases to more than one, with the same cost and service composition. Users' service demand is subject to the two user group regulations, and users choose ISPs and service compositions randomly.

Test analysis: profits and users of the first ISP are shared by other ISPs. Then, this ISP needs to make strategies to attract more users, so as to make up the previous loss.

5.4.3.2 ISPs' Setting Service Compositions

Simulation scene of Test 1: Three ISPs (ISP_1, ISP_2, ISP_3), with the same cost, are likely to offer the following service composition:

$$Plan_i = \begin{pmatrix} 5 & 30 & 1 \\ 10 & 70 & 1 \\ 14 & 100 & 1 \end{pmatrix}, \quad (i = 1, 2, 3)$$

We denote the sum of mobile users by $USER$. Every mobile host, with multiple interfaces, can access more than one ISPs and choose more service compositions.

Process of the Repeated Game: In the first stage, every ISP just provides the first service composition $Plan_i(1, :) = (5, 30, 1)$ ($Plan_i(j, :)$ stands for the j-th row of matrix $Plan_i$); in the second stage, a certain ISP (e.g. ISP_1) provides $Plan_i(2, :) = (10, 70, 1)$; in the third stage, other ISPs offer $Plan_i(2, :) = (10, 70, 1)$ consecutively; in the fourth stage, certain ISPs start to put $Plan_i(3, :) = (20, 150, 1)$ into market; and in the fifth stage, other ISPs provide $Plan_i(3, :) = (20, 150, 1)$ one by one; ...

In Table 5.3, ISP_1 gains more users and profits by providing new service compositions in the second and the fourth stage. Then, the users and profits are shared in the third and the fifth stage. As the sum of user percentage is greater than 1, we know that some users have chosen more than one ISP. Analysis of Test 1: First, for a single ISP, it is appropriate to provide more than one service composition, so as to gain more users and higher profits; second, the total market profits drop in accordance with the increase of service compositions.

It is impossible for ISPs to provide a specific service composition for each user, so there is a need to find the best service composition in the real world. Hence, we should first find the relationship between the chosen threshold of service intervals and characteristics of user groups (average and variance).

In normal distribution, we know that the area of ($\mu - \sigma, \mu + \sigma$) accounts for 68.27 % of the total area, and the area of ($\mu - 1.96\sigma, \mu + 1.96\sigma$) accounts for 95 % of the total area. Hence, we can confirm the threshold of service usage by the density of service demands.

Simulation scene of Test 2: $Plan_1$ of ISP_1 is the same as $Plan_i$ in Test 1. $Plan_2$ of ISP_2 and $Plan_3$ of ISP_3 are made as follows:

Table 5.3 Stage experimental data for the repeated game

Stage	Total profit	ISP_1 profit	ISP_2 profit	ISP_3 profit
1	10.2950	3.5666 (66.9 %)	3.1593 (62.1 %)	3.5690 (68.0 %)
2	9.3756	6.7706 (74.2 %)	1.3141 (20.4 %)	1.2909 (20.1 %)
3	9.3756	3.1489 (37.3 %)	3.1242 (37.5 %)	3.2125 (39.4 %)
4	9.2132	4.5875 (44.2 %)	2.2985 (27.7 %)	2.3273 (28.1 %)
5	9.2132	3.0483 (33.3 %)	3.0974 (33.3 %)	3.0675 (33.3 %)

Table 5.4 Profit comparison of the service plans

Total profit	ISP_1 profit	ISP_2 profit	ISP_3 profit
7.9271	1.0730 (13.7%)	5.0657 (64.3%)	1.7883 (25.5%)

Fig. 5.10 The distribution of amount of service consumed by users of different ISPs

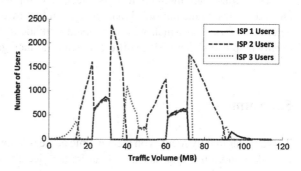

$$Plan_2 = \begin{pmatrix} 4 & 22 & 1 \\ 5 & 30 & 1 \\ 6 & 38 & 1 \\ 8 & 59 & 1 \\ 10 & 70 & 1 \\ 12 & 89 & 1 \end{pmatrix} \quad Plan_3 = \begin{pmatrix} 3 & 14 & 1 \\ 5 & 30 & 1 \\ 7 & 45 & 1 \\ 8 & 50 & 1 \\ 10 & 70 & 1 \\ 13 & 92 & 1 \end{pmatrix}$$

In $Plan_1$, ISP_2's free service usage is the part between $\mu - \sigma$ and $\mu + \sigma$, ISP_2's free service usage is the part between $\mu - 1.96\sigma$ and $\mu + 1.96\sigma$.

Table 5.4 shows the profits of ISP_1, ISP_2 and ISP_3 when they adopt different compositions. Figure 5.10 shows users' choices on ISPs (the number of users is denoted by $USER$).

From Table 5.4, we find that ISP_2, by adopting strategies which are related to the user group characteristic parameter, can get the most profits and users, then ISP_3, and ISP_1 the least.

In Fig. 5.10, users choosing ISP_2 are distributed widely, then ISP_3, and ISP_1 the least.

Analysis of Test 2: Comparing the three ISPs in the tests above, ISP_2 has the most profits and users, because it is made based on the user group parameter (average and variance).

Therefore, ISPs' threshold of service interval should be chosen based on the user group parameter, providing multiple service compositions for different user groups.

5.4.3.3 Summary of the Oligopoly Market

Test 1 has proved ISPs' dynamic game process in the oligopoly market. That is to say, a certain ISP provides new service compositions in every stage, and the Nash equilibrium is that every ISP adjusts their strategies to the same and re-share users and profits. But the total profits will decline in accordance with the increase of the number of games. Under ideal conditions, the purpose of the game is to find the best service composition for every user.

Test 2 illustrates methods to make service compositions based on user group characteristics. In the non-ideal market, we can first investigate the user group characteristics, based on which we can rationally divide user-populated areas, so as to evenly map users' demands to different areas and then gain more users and profits.

5.5 Summary

This chapter mainly focuses on the dynamic game relationship between ISPs (service provider) and multi-interface mobile users (service chooser). We first discuss three types of Internet service markets in an incomplete competition, and build a general service composition model by surveying major ISPs in the market and a service demand model by investigating the need of multi-interface mobile users. Then, we analyze the bargaining game in the exclusive monopoly market and the dynamic repeated game process in the oligopoly market, proving that under ideal conditions, ISPs can get more users and profits if they provide more service compositions different from others'. It also concludes that the subgame perfect Nash equilibrium of the finite repeated game under this condition is that every ISP adjusts their service compositions to make them the same in every stage, resulting in that ISPs provide a specific service composition for each user. Under the specific condition of infinite repeated games, there exists a specific subgame perfect Nash equilibrium, which means all ISPs choose not to adjust their service compositions. Finally, we confirm ISPs' position of price monopoly in the exclusive monopoly market and provide pricing advices for ISPs. It also confirms the dynamic repeated game process of ISPs in the oligopoly market, and offers methods to make service composition according to user group characteristics.

References

1. Altmann, J., Chu, K.: How to charge for network services—flat-rate or usage-based? Comput. Netw. **36**(5/6), 519–531 (2001)
2. Baumol, W.J., Blinder, A.S., Gale, C.L.: Microeconomics: principles and policy. South-Western Educational Publishing, Cincinnati (2001)
3. Bhargava, H., Choudhary, V.: Second-degree price discrimination for information goods under nonlinear utility functions. In: Proceedings of the 34th Annual Hawaii International Conference on System Sciences, p. 6 (2001)
4. Binmore, K., Rubinstein, A., Wolinsky, A.: The Nash bargaining solution in economic modelling. RAND J. Econ. **17**(2), 176–188 (1986)
5. Cao, X.R., Shen, H.X., Milito, R., Wirth, P.: Internet pricing with a game theoretical approach: concepts and examples. IEEE/ACM Trans. Netw. **10**(2), 208–216 (2002)

6. China Telecom Launches Mix & Match Mobile Package: http://www.marbridgeconsulting. com/marbridgedaily/2013-04-12/article/65046/china_telecom_launches_mix_match_ mobile_package

7. Friedman, J.W.: Oligopoly and the theory of games. North-Holland Publisher, Amsterdam (1976)

8. Harrison, A.E.: Productivity, imperfect competition and trade reform: theory and evidence. J. Int. Econ. **36**(1–2), 53–73 (1994)

9. He, H., Xu, K., Liu, Y.: Internet resource pricing models, mechanisms, and methods. Netw. Sci. **1**, 48–66 (2012)

10. Johnson, J.L.: Probability and statistics for computer science. Wiley Online Library, Hoboken (2003)

11. Keon, N., Anandalingam, G.: A new pricing model for competitive telecommunications services using congestion discounts. INFORMS J. Comput. **17**(4), 248–262 (2005)

12. Lilliefors, H.W.: On the kolmogorov-smirnov test for normality with mean and variance unknown. J. Am. Stat. Assoc. **62**(318), 399–402 (1967)

13. Lin, S., Xu, K., Wu, J.P., Wang, N., Zhang, Z., Zhong, Y.F.: Will the three-network convergence happen?—an evolution model based analysis. In: Proceedings of International Conference on Communications in China 2012, pp. 149–154. IEEE (2012)

14. Ma, R.T.B., Chiu, D.M., Lui, J.C.S., Misra, V., Rubenstein, D.: Interconnecting eyeballs to content: a Shapley value perspective on ISP peering and settlement. In: Proceedings of the 3rd International Workshop on Economics of Networked Systems, pp. 61–66. ACM (2008)

15. Ma, R.T.B., Chiu, D.M., Lui, J.C.S., Misra, V., Rubenstein, D.: On cooperative settlement between content, transit and eyeball Internet service providers. IEEE/ACM Trans. Netw. **19**(3), 802–815 (2011)

16. Mo, J., Walrand, J.: Fair end-to-end window-based congestion control. IEEE/ACM Trans. Netw. **8**(5), 556–567 (2000)

17. Notz, W.: International private agreements in the form of cartels, syndicates, and other combinations. J. Polit. Econ. **28**(8), 658–679 (1920)

18. Novshek, W., Sonnenschein, H.: General equilibrium with free entry: a synthetic approach to the theory of perfect competition. J. Econ. Lit. **25**(3), 1281–1306 (1987)

19. Rasmusen, E.: Games and information: an introduction to game theory, 4th edn. Blacwell Publishing, Malden (2007)

20. Valancius, V., Lumezanu, C., Feamster, N., Johari, R., Vazirani, V.V.: How many tiers?: pricing in the Internet transit market. In: Proceedings of SIGCOMM 2011, pp. 194–205. ACM (2011)

21. Wang, Q., Chiu, D.M., Lui, J.C.S.: ISP uplink pricing in a competitive market. In: Proceedings of International Conference on Telecommunications, pp. 1–6 (2008)

22. Xu, K., Zhong, Y.F., He, H.: Can P2P technology benefit ISPs? A cooperative profit-distribution answer. http://arxiv.org/abs/1212.4915

23. Yaïche, H., Mazumdar, R.R., Rosenberg, C.: A game theoretic framework for bandwidth allocation and pricing in broadband networks. IEEE/ACM Trans. Netw. **8**(5), 667–678 (2000)

24. Yuksel, M., Kalyanaraman, S.: Pricing granularity for congestion-sensitive pricing. In: Proceedings of IEEE International Symposium on Computers and Communication 2003, vol. 1, pp. 169–174. IEEE (2003)